Library of Arabic Linguistics

The reasons behind the establishment of this Series on Arabic linguistics are manifold.

First: Arabic linguistics is developing into an increasingly interesting and important subject within the broad field of modern linguistic studies. The subject is now fully recognised in the Universities of the Arabic speaking world and in international linguistic circles, as a subject of great theoretical and descriptive interest and importance.

Second: Arabic linguistics is reaching a mature stage in its development benefiting both from early Arabic linguistic scholarship and modern techniques of general linguistics and related disciplines.

Third: The scope of this discipline is wide and varied, covering diverse areas such as Arabic phonetics, phonology and grammar, Arabic psycholinguistics, Arabic dialectology, Arabic lexicography and lexicology, Arabic sociolinguistics, the teaching and learning of Arabic as a first, second, or foreign language, communications, semiotics, terminology, translation, machine translation, Arabic computational linguistics, history of Arabic linguistics, etc.

Viewed against this background, Arabic linguists may be defined as: the scientific investigation and study of the Arabic language in all its aspects. This embraces the descriptive, comparative and historical aspects of the language. It also concerns itself with the classical form as well as the Modern and contemporary standard forms and their dialects. Moreover, it attempts to study the language in the appropriate regional, social and cultural settings.

It is hoped that the Series will devote itself to all issues of Arabic linguistics in all its manifestations on both the theoretical and applied levels. The results of these studies will also be of use in the field of linguistics in general, as well as related subjects.

Although a number of works have appeared independently or within series, yet there is no platform designed specifically for this subject. This Series is being started to fill this gap in the linguistic field. It will be devoted to Monographs written in either English or Arabic, or both, for the benefit of wider circles of readership.

All these reasons justify the establishment of a new forum which is devoted to all areas of Arabic linguistic studies. It is also hoped that this Series will be of interest not only to students and researchers in Arabic linguistics but also to students and scholars of other disciplines who are looking for information of theoretical, practical or pragmatic interest.

The Series Editors

A linguistic study of the development of scientific vocabulary in Standard Arabic

Abdul Sahib Mehdi Ali
Head of the Department of Translation
al-Mustansiriyyah University, Baghdad

A linguistic study of the development of scientific vocabulary in Standard Arabic

Monograph No.6

Routledge
Taylor & Francis Group
LONDON AND NEW YORK

First published 1987 by Kegan Paul International

2 Park Square, Milton Park, Abingdon, Oxon OX14 4RN
711 Third Avenue, New York, NY 10017, USA

Routledge is an imprint of the Taylor & Francis Group, an informa business

First issued in paperback 2016

Transferred to Digital Printing 2010

British Library Cataloguing in Publication Data
A catalogue record for this book is available from the British Library

ISBN 978-1-138-99543-7 (pbk)
ISBN 978-0-7103-0023-2 (hbk)

Publisher's Note
The publisher has gone to great lengths to ensure the quality of this reprint but points out that some imperfections in the original copies may be apparent. The publisher has made every effort to contact original copyright holders and would welcome correspondence from those they have been unable to trace.

To the soul of my father who was
my constant inspiration in seeking
knowledge and appreciating it; and
to my family, whose unfailing
support has been invaluable to me.

Editor's note

The subject of this book is one dear to the heart of Arab scholars, but which may raise matters completely unfamiliar to English readers coming from a linguistic background characterized by a somewhat *laissez-faire* towards the adoption of foreign vocabulary. English is a hybrid language, basically Germanic, with a 900 year history of absorption of Romance vocabulary, first through Norman French and then as a result of scholarly borrowings and neologisms from Latin. It contains an immense proportion of foreign vocabulary regarded, completely without embarrassment, as English. Arabic, on the other hand, is in a completely different situation. It can, of course, no more claim to be totally homogenous in its vocabulary than any other language can. Nevertheless, the overwhelming proportion of its core vocabulary is native or else the result of loans of such antiquity as to be indistinguishable from native vocabulary.

The various colloquial dialects have, it is true, absorbed foreign vocabulary from the foreign cultures (both western and eastern) with which they have been in contact. However, the scholarly establishment of the Arab world has always resisted the absorption of this into the written language. With the recent technological and educational developments which have taken place in the Arab world, a need has arisen for the transfer of technical concepts into Arabic in many fields. Some scholars have felt that the only acceptable method for this is to extend the meanings of existing Arabic words or to coin new words through the process of derivation from existing roots. This technique is known as *Qiyās* – derivation by analogical formation. Others have been inclined to absorb foreign vocabulary, but modify this to fit Arabic phonological and, if possible, morphological patterns. This is known as *Taʿrīb* – Arabization.

The Arabs are justly proud of the richness of their vocabulary and the great number of synonyms and words expressing minute differences of meaning for a basic lexical concept. The proponents of the technique of *Qiyās* felt that the existence of such a great number of roots, coupled with the Semitic morphological techniques of forming related lexical concepts by affixing or infixing, could with a certain amount of ingenuity cope with any needs for added vocabulary. In addition to this, it was felt that archaic words, no longer widely used, could be employed for new concepts. The supporters of Arabization, on the other hand, took the view that although coining from Arabic roots was preferable where feasible, the obscureness of some of these coinings was so extreme that it would represent a great saving of effort to take over foreign words, which could have the virtue of familiarity to those already educated in foreign languages, and would form a link between western and eastern cultures.

Ali examines both these techniques and attempts to give guidelines for future development. His results derive from theoretical examination of the subject, consultation with Arab scholars, examination of existing technical lexicons and journals, and on a survey made with students of technical subjects aimed at measuring their readiness to accept neologisms formed in these different ways and the ease with which they could understand them. He has also explored the different techniques of word formation employed, including "blending", "fusion" and "compounding", and reports on the reaction of students to words produced in these ways.

Ali's work represents a new stage in Arab scholarship which combines a respect for tradition based on a thorough knowledge of the language and traditional linguistic writing, coupled with an open-minded approach to investigation of the actual problems faced by students and the way in which vocabulary is used in technical writing today.

Bruce Ingham
Series Co-Editor

Contents

TABLES

Acknowledgements

I would like to express my deepest gratitude to Dr Bruce Ingham for taking the responsibility of supervising the work on my doctoral thesis, upon which this book is based. His unfailing patience, his friendly attitude, and his deep insights in the language have been invaluable. I would also like to thank Dr R. Hayward and Dr T. Bynon of the School of Oriental and African Studies for the kind help they offered me.

I am grateful to all my friends and colleagues with whom I discussed various aspects relating to Arabic scientific vocabulary, and whose remarks have been most useful and enlightening. Special thanks are due to Dr Yahya A.M. Ahmad, Dr Abdul-Emir al-Saffar, W. Abderrahman, Khalid A.S. Muhammad, Abdullah al-Dawsari and Peter Colvin. Thanks are also due to Salim H. Salih, of al-Hashimiyyah Secondary School, Babylon, for help with the questionnaire.

I also wish to express my thanks to Dr Ali M. al-Kasimi, of the Permanent Bureau of Co-ordination of Arabization, and Ahmad A. Ghazal, General Director of the Institute of Researches and Studies of Arabization, both at Rabat, Morocco, who gave me much of their valuable time during my fieldwork visit to their organizations.

I am greatly indebted to the Iraqi Ministry of Higher Education and Scientific Research for granting me the necessary scholarship funding whereby I was able to read for the degree of PhD and do the work on which this book is based.

Finally, I would like to express my sincere thanks to my family, who have always overwhelmed me with their love, kindenss and support.

Transcription and transliteration systems

The transcription system adopted in this book is intended to give the Arabic examples a reading form; hence it is a reading transcription and not a strictly phonetic one. The following list of symbols represents the Arabic phonological system:

Consonants

b voiced bilabial plosive

m voiced bilabial nasal

w voiced labio-velar semi-vowel

f voiceless labio-dental fricative

ṯ voiceless interdental fricative

ḏ voiced interdental fricative

z voiced alveolar fricative

s voiceless alveolar fricative

t voiceless denti-alveolar plosive

d voiced denti-alveolar plosive

n voiced alveolar nasal

r voiced apical trill

l voiced apical lateral

y voiced palatal semi-vowel

j voiced alveo-palatal affricate

š voiceless alveo-palatal fricative

k voiceless velar plosive

q voiceless uvular plosive

x voiceless uvular fricative

ġ voiced uvular fricative

ḥ voiceless pharyngeal fricative

ᶜ voiced pharyngeal fricative

? glottal stop

h voiceless glottal fricative

Emphatics

Arabic has a set of so-called "emphatic" sounds: *ṭ ṣ ḍ ḏ̣* which correspond to the above unemphatic ones: *t s d ḏ* and which differ from the latter in that their articulation involves velarization, i.e. added raising of the back of the tongue towards the soft palate, a phenomenon known in Arabic by the name *?iṭba:q*. The following is a brief description of these sounds:

ṭ voiceless denti-alveolar plosive (velarized)

ṣ voiceless alveolar fricative (velarized)

ḍ voiced denti-alveolar plosive (velarized)

ḏ̣ voiced interdental fricative (velarized)

Vowels

i short, front, unrounded vowel between close and half-open

i: long, front, unrounded, close vowel

a short, open, front vowel

a: long, open front vowel

u short, back, rounded vowel between close and half-open

u: long, close, rounded back vowel

Other sounds

In cases where reference is made to Colloquial Arabic or loanforms that have not undergone full phonological assimilation, the following symbols will be used to represent phonemes which do not exist in Standard Arabic:

p voiceless bilabial plosive

v voiced labio-dental fricative

č voiceless palato-alveolar affricate

g voiced velar plosive

e short, half-open, front unrounded vowel

o short, half-open, back, rounded vowel

e: half-close, front, unrounded long vowel

o: half-close, back, rounded, long vowel

Other notational hints

1 A long vowel is marked by two dots; a long consonant by doubling the symbol, e.g.

katab-	"to write"	*ka:tab-*	"to correspond"
daras-	"to study"	*darras-*	"to teach"

2 In this transcription system the type of representation a word has depends on whether it is used in a context or as a citation form. In the former case, words are represented with their case-endings, except those which occur finally, i.e. in pre-pausal form. When used in isolation, on the other hand, words are deprived of their inflections. For example:

In isolation:	In context:
našа:ṭ "activity"	*naša:ṭun ?iš^c a:^c iyy* "radioactivity"
	naša:ṭun ?iš^c a:^c iyyun ṭabi:^c iyy "natural radioactivity"

The same applies in the case of quotations and sentences. In the case of verbs used in isolation, as a general rule the exclusion of inflections is indicated by a dash following the element which constitutes the base or root of the verbal unit, e.g. *kasar-* "to break". In cases where the retention of these inflections is found necessary, the dash is still used to separate the two elements, e.g. *kasar-tu* "I broke (something)". On the other hand, when used in a context, the verbal unit is fully represented and without the use of a dash, e.g. *?ana: kasartu l-qalam* "I broke the pen".

3 The definite article *?al-*. When attached to a word used in isolation, the initial glottal stop of this article will not be represented, and the remaining *al-* will be separated from the word it defines by a dash, e.g. *al-kita:b* "the book". When prefixed to a word beginning with one of the letters:

t ṯ d ḏ r z s š ṣ ḍ ṭ ḏ̣ l n

the *l-* of the definite article is assimilated to the sound it is annexed to, e.g.

al- + ṯalj	*aṯ-ṯalj*	"the snow"
al- + sama:?	*as-sama:?*	"the sky"
al- + ṭayr	*aṭ-ṭayr*	"the bird"

In junctural position the definite article may be represented by the letter *l-* or the sound it assimilates to only, as in:

miqya:su l-ġalaya:n	"hypsometer"
quwwatu š-šira:?	"purchasing power"

It should be noted that the attachment of the definite article to the word it defines, as in the examples just given, is done on purely grammatical grounds; otherwise this involves a violation of the rules of Arabic syllabic structure. For, according to one of these rules, the permissible types of syllabic structures in junctural position are only CV, CV, and CVC. To take the first of the two examples above, the article *l* goes with the preceding word *miqya:su* to yield the allowable syllables CVC, CV, and CVC. Grammatically speaking, however, *l* is attached to *ġalaya:n*, thus producing the structure *l-ġalaya:n* CCV-, which is not in keeping with the aforementioned rule.

4 In pre-pausal position the feminine marker *-at* becomes *-ah*, e.g. *madrasatu l-bani:n* "school of boys", *madrasah* "school".

5 The letters AH after a date indicate that the dating is in the Muslim era, which begins with the emigration (*hijrah*) of the Prophet Muhammad from Mecca to Medina in 622 AD. The absence of the foregoing letters indicates that the time division falls within the Christian era, though the initials AD (Anno Domini) may also be used for the same purpose.

Transliteration

The transliteration system used is one commonly used for Arabic and is employed here for the notation of proper names, terms and titles of works, etc. and employs the digraphs:

	th	dh	ḍh	sh	gh	kh
instead of	ṯ	ḏ	ḏ̣	š	ġ	x

It also employs the macron for length, giving

	ī	ā	ū	ō
instead of	i:	a:	u:	o:

As further simplifications and for the sake of typographical elegance, the glottal stop ? is omitted in initial position to give Adab instead of ?adab, and the assimilations of the /l/ of the article preceding "sun letters" are not marked, giving al-naḥwiyyīn (not an-naḥwiyyīn).

Introduction

The problem and its dimensions

After a long period of stagnation, starting from the end of the thirteenth until the nineteenth century – what may be described as an age of decadence – a movement of linguistic renaissance began to take place in the Arab world, beginning in Egypt, spreading to Lebanon, and thence to the rest of the Arab nation. This movement represented one aspect of a multidimensional process of revival and revitalization which the nineteenth-century Arab society came to experience as a result of its awakening to the realization that it had to face the challenge of European culture and civilization, with which it was gradually coming into contact.

One of the important developments which helped create that contact was the introduction of the Arabic printing-press in Egypt during the Napoleonic expedition in 1798.(1) Through the dissemination of printed works, newspapers, magazines, etc., knowledge of the cultural, scientific and technological contributions of Western civilization gradually came to be within the reach of the Arabs in the various sectors of their world. The new discoveries, ideas, concepts, etc. of modern European civilization were also made accessible to the Arab world through the cultural, educational and scientific missions which were sent to Europe for the acquisition of scientific knowledge and specialized training in various disciplines: medicine, natural sciences, economics, law, administration etc. On returning home, scholars engaged in such missions transmitted their acquired knowledge and experience to their fellow-countrymen via a variety of channels: writing books, articles, treatises, etc. or translating them;(2) working as teachers or instructors in schools, institutes, universities, etc.; establishing or participating in the establishment of academic, social, cultural, or scientific organizations, and so on.

An important aspect of this cultural confrontation between the East and the West is that, to the Arabs, it represented a strong jerk, so to speak, which offered them the opportunity to rediscover themselves, to consider the state of affairs they had been reduced to after the decline of their empire, to remind themselves of their glorious past as a nation of considerable cultural weight, as contributors to world civilization rather than merely recipients from or dependants on it.

One significant development resulting from the Arabs' awakening to this situation was the emergence in the Arab world of a large number of nationalist, political, religious and linguistic reformers and leaders whose great admiration for their cultural heritage inspired them with the feeling that their nation, too, had its honourable achievements and great potential for further development. To mention only a few of those reformers and leaders, there were: Muḥammad Ali (1769-1849), Nāṣīf al-Yāziji (1800-1871), Rafā'ah al-Ṭahṭāwi (1801-1873), Maḥmūd Shukri al-Ālūsi (1802-1854), Aḥmad Fāris al-Shidyāq (1804-1887), Buṭrus al-Bustāni (1819-1893), Jamāl al-Dīn al-Afghāni (1839-1897), Ibrāhīm al-Yāziji (1847-1906), Muḥammad Abduh (1849-1905), Sulaymān al-Bustāni, and many others.(3)

The Arab world also witnessed the birth of many organizations – social, political, cultural, scientific – which played an increasingly important role in the enlightenment of society and the spread of knowledge. Similar contributions were also made by universities, schools, libraries, and other educational institutions which were being established in many parts of the Arab world.

The influence of the circumstances and developments discussed above soon came to embody itself in the persistent calls on the part of numerous leaders, reformers and writers, advocating social and cultural changes in Arab society to help it resume its growth and meet the challenges of modern world civilization. Although attitudes regarding the way these objectives might be attained could not be said to have been unanimous, unanimity was not lacking in respect of one important ultimate goal, namely, the preservation of Arab society.(4) The precedence taken by this goal is accounted for by the reasoning on the part of modern Arabs that for their nation to be able again to assume a cultural status similar to that it enjoyed in medieval times, it has to retain its original character and protect its national integrity.

Modern Arabs have also learned that the preservation of their society requires their unity, that the decline of their empire had been primarily due to the state of disintegration which had penetrated the body of their nation. Hence the dream of modern nationalist movements to establish a unified Arab state and abolish the boundaries and differences now existing in the Arab world. An important fact to be noted in this respect is that there has been a growing awareness among the Arabs in their various countries that the Arabic language, being the mother tongue of the whole nation and the container of its culture, represents a unifying factor of utmost importance, and that any attempt at unification which fails to give this fact its due weight is bound to fall short. It is therefore an agreed principle that for the language to play this role, as it did in medieval times, and be a vehicle of culture and science as well as religion, it must not only be revitalized and enriched, but also unified and standardized. In this way Arabic is seen as a fundamental bond providing Arab people all over their world with a sense of unity and identity.

Efforts aimed at unifying and standardizing the language have been on the increase since the early decades of the present century.(5) Individuals as well as language academies have been well aware of the fact that unless their efforts at reviving the language were fully co-ordinated, they would fail to be of any significance at the nationwide level. Without such co-ordination, it would not be surprising, for instance, to find a multiplicity of terms provided by different Arab countries for one and the same (newly introduced) object and concept. This state of affairs is likely to lead to confusion. As one Arab writer has rightly observed: "the existence of a profusion of synonyms is no longer regarded as a sign of linguistic richness or as a reflection of the inherent quality of a language" (Ibn Abdallah, 1976, p.vi). This being the case, lack of co-ordination of efforts is generally seen as an urgent problem in its own right.

A serious and wide-reaching attempt at unifying the Arabs' efforts to revive their language, and especially to standardize its scientific and technical vocabulary, can be said to have been made with the establishment of the Arab League in 1945, which had under its auspices a cultural committee concerned with the general task of promoting cultural unity in the Arab world. The year 1946 witnessed the conclusion among the Arab countries of a cultural agreement which defined its position regarding the standardization of the language and enhancing of its role as a medium of scientific expression as follows (quoted in the "Middle East Journal", vol.i, January 1947, p.208):

> The states of the Arab League will work for the standardization of scientific terms, by means of councils, congresses and joint committees, which they will set up and by means of bulletins which these organizations will issue. They will work to make the Arabic language convey all expressions of thought and modern science, and to make it the language of instruction in all subjects and in all educational stages in the Arab countries.

It should be mentioned, however, that these and other efforts intended for the same objective, that is, linguistic standardization and specifically terminological unification, have not yet yielded the desired results. Language academies in Damascus, Cairo and Baghdad have not so far attained an appreciable level of success in co-ordinating their contributions. Hence the early realization on the part of Arab educational circles that a Pan-Arab academy of the language should be established (this call was raised at a meeting of Arab ministers of education in Cairo in 1953). In response to the preceding call, a conference was held in Damascus in 1956 under the name "Conference of the Arab Language and Science Academies", which aimed at enhancing the process of co-ordination among the various academies.(6) The same call was again echoed at a Pan-Arab conference on Arabization held in the Moroccan capital, Rabat, in April 1961. Calls for the establishment of a unified academy of the Arabic language have continued unabated up to the present moment, and the need for such an academy is becoming more and more pressing with the passage of time.

The reader will have inferred from the above remarks that full co-ordination among Arab language academies is still an unattained objective. However, this is not the only difficulty, perhaps not even the most serious one facing the language in its present stage. What may be rightly described as *THE* problem of present-day Arabic, is the considerable lack of scientific and technical terminology in most branches of science and technology — so much so that a large number of institutions of higher education in various Arab countries still resort to a foreign language (English or French) as a medium of instruction for science. This problem can be accounted for mainly in terms of the events and circumstances experienced by the Arab world up to the contemporary awakening, which some date as beginning not earlier than the second half of this century. Prior to this, during the Ottoman and Western colonization of the Arab world, the Arabic language was in general denied the opportunity of being used as the medium of instruction in its homeland, nor was it the language of administration. Arabic was also under the disadvantage that in the period preceding this awakening, due to the impact of many adversities, the Arab world had been brought to such a state of apathy and paralysis that little or no initiative was shown in the field of scientific productivity. And since scientific terms are usually created by scientists and in scientific laboratories, the absence of such circumstances was one of the main factors accounting for the non-existence or paucity of such coinages in that period.

In addition to the gulf of several centuries of inactive scientific life, there is also the problem that (especially nowadays) "the great number of scientific and technical terms that are produced every day in the industrial and post-industrial countries makes it very difficult for Arabic to catch up" (al-Kasimi, 1978, p.15). According to Ghazāl (n.d., p.10) every year there are coined about 7300 terms relating to the various branches of science; in Arabic, however, this number goes down to 2500 terms (coined by organizations, committees, scholars).

The problem of neologization and lexical growth in scientific Arabic involves a large number of other aspects of intra-, as well as inter-lingual nature, plus various types of extra-lingual factors. These are discussed in full detail throughout the body of this book; meanwhile, we may as well provide the reader with a brief statement justifying the purpose for which we have undertaken the present task.

Purpose and justification

The problem referred to above, despite its seriousness and the fact that it has engaged the attention of so many scholars, linguistic bodies as well as governmental institutions, has unfortunately hardly been the subject of an objectively oriented and scientifically based approach where facts and findings of modern linguistic studies form the basic guidelines along the way to the ultimate truth. Attempts at discussing certain aspects of this problem have not been lacking, as we shall see, but they have generally failed to attain the required level of accuracy, adequacy and/or objectivity. This analysis is therefore intended to avoid the foregoing shortcomings and reach the standard of appropriateness aspired to by the writer. With this view in mind, it is hoped that this attempt will provide a better opportunity for seeing matters in their right perspective. The fulfilment of this objective will help to shed some light on the various aspects of the problem and thus render us more capable of accounting for the different phenomena involved.

Our main objective is to study the development of scientific vocabulary in Arabic throughout its history, with special emphasis on the phenomenon in modern times. We also aim to examine and assess the capability of the language to cope with the ever-increasing demands of human civilization, scientific and technological advances, and social and cultural developments. Another objective is to underline and highlight the potential of Arabic for self-enrichment, and its natural tendencies for growth and development. This, of course, entails a careful study of the morphological properties and word-formational processes characteristic of the language.

At a different level, another objective is to examine critically various contributions, ideas, proposals, etc. which have been put forward by a large number of writers, academicians and lexicographers in an attempt on their part to provide what (to them) constitutes appropriate measures prompted by certain terminological needs. This critical account attempts to evaluate matters pertaining to their validity, practicability, and conformity to the spirit of the language. To substantiate the views and arguments presented, we have as a rule turned to actual linguistic practice for evidence to serve our purpose. In many cases supporting evidence is also drawn from other languages where what is considered with regard to Arabic is also applicable, i.e. in the case of linguistic, sociolinguistic and cultural characteristics which have more or less universal validity.

Besides the native resources and techniques of lexical expansion available to Arabic (derivation, revival of the vocabulary, semantic extension, etc.) there is another but non-natively based method, namely Arabicization (incorporation of foreign elements). This has always been a most controversial issue among Arab scholars and in linguistic circles – so much so that it has been found essential to dedicate a part of this book (Chapter 4) to the discussion of this subject and the various linguistic and extralinguistic (scientific, cultural and nationalistic) aspects associated with it. By so doing, it is hoped to study the significance of borrowing as a means of expanding Arabic scientific terminology, and to evaluate the role it has played in the configuration of the language over several stages of its history.

This introduction touches upon only the most important objectives of the book. At the beginning of each chapter, there is also a brief account of the aims and objectives underlying the discussions therein.

Notes to Introduction

1 See Zaydān, "al-Lughah" (n.d.), p.106.
2 A typical example in this connection is Rifā[c]ah al-Ṭahṭāwiyy 1801 - 1873). He, when twenty-five, travelled to Paris at the head of a cultural mission sent by Muḥammad Ali. An interesting account of his impressions of Parisian society and the impact of its cultural accomplishments on his thinking is well worth reading in his book "Takhlīṣ al-Ibrīz"(see below, p.110).
3 For a concise account of these and other personalities and their contributions to Arab revivalism, see Chejne (1969), pp.85-144.
4 Ibid., p.92.
5 The earliest attempt in this respect was represented by the convening of an Arab Congress of Physicians in Baghdad in 1938, which dealt with the formulation and standardization of medical terms. As we shall see, similar attempts have continued ever since.
6 For the proceedings of this congress,(Mu?tamar al-Majāmi[c] al-Lughawiyyah al-[c]Arabiyyah), see MM[c]I[c]A, vol.xxxii, no.1, 1957.

Language

The following terms are used to designate the various varieties of the Arabic language treated in this book:

Classical Arabic (CA)

This term is used to refer to the Classical Arabic of medieval times, the language of the Qur?ān and the vehicle of pre-Islamic poetry, in which a wealth of literature is written. Following its codification in the early years of Islam, it gradually acquired its standard form and established itself as the language of science and administration.

Standard Arabic (SA)

In default of a more appropriate term, Standard Arabic is used here to designate the variety of Arabic common to the whole Arabic-speaking world and which constitutes the normal medium for all written communication.

(This variety has also been designated by a number of other terms which, however, seem to be less satisfactory. The term "Contemporary Arabic", for instance, fails to exclude the colloquial varieties of present-day Arabic; the same is also true in the case of "Modern Arabic". "Literary Arabic" is too specific to include manifestations which have nothing to do with literature, newspapers and advertisements; and "Written Arabic" fails to suggest that the variety in question is also used as a medium for spoken communication. "Standard Arabic", which does not give rise to any serious criticism, may thus be said to be the least objectionable of these terms.)

SA is based on and inspired by CA, and although it has developed and acquired new vocabulary, it has kept in line with the characteristic morphological, grammatical and syntactic properties of the latter. It is found in contemporary books, documents, newspapers, magazines etc.; and orally it is the vehicle for formal speeches, public lectures, educated conversations, learned debates and news broadcasts over radio and television. Like CA, SA is characterized by being the language of the educated sectors of society and is regarded by them as being of far greater prestige than dialectal varieties. For the sake of brevity, the term Arabic will often be used unqualified to refer to Standard Arabic.

Colloquial Arabic or Dialectal Arabic

This term is occasionally used to refer to the range of non-standard localized varieties of Arabic employed in the speech of everyday life throughout the Arabic-speaking countries. These are almost entirely spoken forms and each of them may be characterized by certain features unique to it and others of a wider application (i.e. exhibited by more than one dialect). Where reference to a particular dialect of a particular country is necessary, a qualifying term will be used to specify that dialect. For example, "Iraqi Dialect" or "Iraqi Arabic" will refer to the type of colloquial language used in Iraq.

1 Arabic word structure

One of the most important convictions of the modern linguist is that every human language is a system which involves a great degree of complexity, far greater than we normally realize. It has also been established that what applies in the case of one language may not necessarily apply in another. In view of these facts and due to the nature of our study (in which various aspects of linguistic structure characteristic of more than one language are frequently discussed and compared) we have found it necessary to supply the reader with a clear idea of the fundamental principles of Arabic word structure. Another important purpose to be served by the present discussion consists in the realization that for our analysis of the various issues to be precise and scientific, it must employ a set of clearly defined concepts in accordance with which facts and claims can safely and confidently be stated.

However, we do not propose to discuss here all the possible issues that may be included under the heading Arabic Word Structure, as some of these are treated elsewhere in this book where it has been deemed more appropriate.(1) In the present chapter priority has been given to those facts and pieces of information which are necessary for paving the way towards clearer understanding of the problems discussed, as well as a precise and consistent account of their analysis.

1.1 The Word

Native speakers of Arabic, like those of any other language, seem to be intuitively aware of what the term "word" means in its general sense, and it can safely be argued that this awareness applies both to speakers who write the language as well as those who do not. But what about "word" as a linguistic unit? This question has given rise to much controversy in linguistic circles. The controversy has been such that in spite of the fact that much ink has been spilt, no conclusive definition of what a word is or how it should be defined has been achieved. One linguist, Malinowski, has gone so far as to say that words "are in fact only linguistic figments".(2) Bloomfield's definition of word as "a minimum free form",(3) which is widely celebrated, has not been immune from criticism, as it fails to account for the fact that many words (e.g. "the", "my", or French "je", etc.) can hardly stand alone.(4) Recognizing the difficulties with the definition, Palmer (like many other linguists faced with the same situation) cannot help concluding his section on "words" by stating: "sadly, we have to say that the word is not a clearly definable linguistic unit" (Palmer, 1971, p.51).

As a result of these difficulties, there now exist several approaches to the problem, which represents a tendency on the part of linguists to distinguish the concept of word on several levels. It may thus be seen as a phonetic or phonological unit (one that is bound by pauses), as a morphemic unit (defined in its position in the sentence), as a semantic unit with a specific meaning, etc.

The discrepancies and difficulties involved in giving a direct definition of word, and which have been experienced by modern linguists in Europe and America, have long revealed themselves in the writings of early Arab grammarians. Their definitions, although based on the nature of the Arabic language,(5) have similarly proved to be open to criticism on various grounds.(6) We need not go into further details, since we have no intention of adding fuel to an already well-stoked controversy. In this book we will employ the morphological word as the basic lexical unit of citation. The great majority of our words are given in pausal form; there are occasions, however, where contextual forms are also cited by way of exemplifying or demonstrating the cases or reasons for their occurrence in such forms. The appendix contains a number of various types of neologisms quoted from different sources and cited in the contexts in which they originally appeared.

1.2 Standard Arabic Stems

Let us take for our point of departure the following definition of word: A word is any free form which consists of a single stem with or without accompanying affixes. In the light of this definition an Arabic word can be said to be represented by any of the following:

(a) a stem alone, e.g. *ru:ḥ* "soul, spirit"
(b) a stem + one or more affixes, e.g. *ru:ḥiyy* "spiritual", *ru:ḥiyyah* "spirituality"
(c) a stem always used alone, i.e. never with an affix, e.g. *ṯumma* "then"

The above types of stems are called Free Stems, as opposed to another type, Bound Stems, where the stem never occurs without an affix. An example of this latter type is represented by the element *-ḥra:r* in *miḥra:r* "thermometer" and *taḥra:r* "recalescence".(7)

A further distinction may also be made between Complex and Simple (or Simplex) Stems, i.e. if we look at them from the point of view of their internal structure. If we go back to the examples cited in (b) above, we find that a word like *ru:ḥiyyah* constitutes a complex stem analysable into the stem *ru:ḥiyy* and the derivational suffix *-ah*, which indicates abstract nouns. The stem *ru:ḥiyy-* itself is a complex one capable of further analysis into the simple stem *ru:ḥ* and the adjectival (derivational) suffix *-iyy* (indicating reference). With regard to the stem *ru:ḥ*, our analysis cannot be carried out any further; hence its name Simple Stem. Thus a Complex Stem is that which involves the combination of a stem plus one or more affixes and to which other affixes may still be added.

1.2.1 Simple and Complex Stems

One of the conspicuous characteristics of Arabic simple stems is their divisibility into two major constituents, a radical morpheme, the so-called root, and an interradical vocalic one, sometimes called non-root. The former is the bearer of the general lexical meaning, the latter is mainly derivational or grammatical. The combination of these two morphemes yields the stems. For instance, the root s-b-k is associated with the general meaning of "founding metal"; this meaning is modified in various ways by the various types of vocalic morphemes in combination with which the root is used. The foregoing root may thus yield the following meanings:

Radical Morpheme (root)	+	Vocalic Morpheme (non-root)		
		-a-a-*	*sabak-a*	he founded
s-b-k		-u-i-	*subik-a*	it was (has been) founded
		-a--	*sabk*	founding
		etc.	etc.	

*The symbol "-" is used here to indicate the place of a root consonant. Notice that in the case of root symbolization, as in s-b-k, its function is to separate the root consonants.

Notice that the same vocalic morphemes are capable of occurrence with different roots, producing similar functional meanings, e.g.

Root	General Meaning	-a-a-	-u-i-	-a--
k-s-r	to break	*kasar-*	*kusir-*	*kasr*
?-k-l	to eat	*?akal-*	*?ukil-*	*?akl*
ḍ-r-b	to hit	*ḍarab-*	*ḍurib-*	*ḍarb*
f-t-ḥ	to open	*fataḥ-*	*futiḥ-*	*fatḥ*
m-s-k	to catch	*masak-*	*musik-*	*mask*

An important point to be noted here is that the root consonants maintain the same order, relative to each other, in all the derived stems. This point will acquire greater significance when we later discuss other types of derivatives following the same principle.

The above process of root-and-non-root combination gives a limited number of derivatives. In order to meet its need for further forms, the language employs other morphemes which may not be purely vocalic but consist of a sequence of consonant(s) and vowel(s). In this case the additional morpheme(s) may take the form of a prefix, infix and/or suffix. There are cases where it may be described as a discontinuous morpheme. Stems containing such affixes (as has been pointed out) would be described as complex ones. The affix in a complex stem may be one of the following:

(a) A prefix, e.g. *mi-* as in *milqaṭ* "forceps" <l-q-ṭ
(b) An infix, e.g. *-k-* as in *rakkab-* "to synthesize" <r-k-b
(c) A suffix, e.g. *-iyy* as in *falakiyy* "orbital" <f-l-k
(d) A combination of (a) and (b), e.g. *ta-* and *-s-* as in *tarassub* "sedimentation" <r-s-b
(e) A combination of (a) and (c), e.g. *?in-* and *-iyy* as in *?inkisa:riyy* "refractive" <k-s-r
(f) A combination of (a), (b) and (c), e.g. *ta-*, *-r-* and *-iyyah* as in *taḥarrukiyyah* "mobility" <ḥ-r-k

1.3 Lexical Patterns

The reader will have by now realized that the co-occurrence of the root and non-root constituents in the Arabic word gives the resultant form a specific pattern(8) which may be represented by a large number of words relating to different roots. An investigation of Arabic vocabulary would result in the realization that the process of combining radical and non-radical elements into lexical forms follows a set of fairly systematic rules in accordance with which derivatives of the same root may be grouped together to form a family of words, each with a specific type of pattern characteristic of a certain type of function. Examples illustrating this are provided in 1.3.2; meanwhile we will turn to the way roots and patterns are symbolized in this book.

1.3.1 Roots and Patterns Symbolized

The radical elements constituting specific roots are given as sets of letters, usually followed by the general idea or meaning represented by the root, e.g. k-t-b "to write". The same also applies where the root-constituting elements are not all strong radicals (as in the foregoing example), but also include one or more weak ones, e.g. w-z-n "to weigh", s-y-r "to walk", w-q-y "to protect", etc. Roots that are abstracted from loanforms are symbolized in the same way, the only difference being that these are enclosed within quotation marks to distinguish them from native ones. Such cases are represented by "s-f-l-t" "to asphalt", "t-l-f-z" "to televize", "?-k-s-d" "to oxidize", etc.

For the symbolization of patterns we shall use F, 8, and L to indicate, respectively, the first, second and third radicals constituting the triliteral root. Interradical vocalic elements will be represented as they are. For instance, words like *lahab* "flame", *jism* "body", *kusu:f* "eclipse", etc. would be symbolized as Fa8aL, Fi8L, and Fu8u:L respectively. These patterns may be described as simple ones, implying that they consist solely of the radical elements with one or more vowels distributed among them. In cases where the number of radicals in the given stem exceeds three, or where the stem is a complex one, the following points should be observed:

1 If the root contains four radical letters, the fourth one is represented by a repetition of the same symbol used for the third radical in triliteral roots: L. For example a word like *su*^{c}*lu:k* "robber", where the root consists of the radicals

ṣ-^{c}-l-k, would be assigned the pattern Fu8Lu:L.

2 If the additional element represents a doubling of one of the radicals constituting the triliteral root, the symbol used for that doubled radical is also doubled. For example, *rassa:m* "painter", which derives from r-s-m "to paint", has the pattern Fa88a:L.

3 Additional elements or morphemes that do not form part of the radical root are retained in the pattern but represented by small rather than capital letters.(9) Here are some examples:

mijhar	microscope	miF8aL	< j-h-r	to show
?inkisa:r	refraction	?inFi8a:L	< k-s-r	to break
taka:ṯuf	condensation	taFa:8uL	< k-ṯ-f	to become dense
?istibda:l	substitution	?istiF8a:L	< b-d-l	to substitute
mumaġnaṭ	magnetized	muFa8LaL	<"m-ġ-n-ṭ"	to magnetize
tajammud	solidification	taFa88ul	< j-m-d	to solidify

In a number of cases where our main concern is to refer to the general consonant-vowel structure of a word, a capital C will be used to indicate any consonant regardless of whether that consonant is a radical or (part of) a non-radical element. For instance, the words *zilza:l* "earthquake" and *minša:r* "a saw", respectively, relate to the quadriliteral z-l-z-l "to shake" with the pattern Fi8La:L, and the triliteral n-š-r "to saw apart" with the pattern miF8a:L. However, both can be described as having the general structure CiCCa:C.

Non-analogical loanforms, foreign borrowings whose morphological structure is inconsistent with that of Arabic words (see Chapter 4), are usually not treated in terms of the system explained above. These are forms like *?ilikitro:n* "electron", *mikrosko:b* "microscope", *helikoptar* "helicopter", etc.(10) However, by way of indicating the non-nativeness of certain patterns, or contrasting them with native ones, we have on occasion symbolized such unassimilated patterns by using an asterisk to the left of the pattern to distinguish them from native ones. On the other hand, analogical loanforms, i.e. those that have been brought into line with the Arabic system, are assigned the appropriate Arabic patterns. For example: *film* <Eng./Fr. "film" would be represented by Fi8L, *faylasu:f* <Gr. φιλόσοφος "philosopher" by Fay8aLu:L, *sirda:b* <Per. sard-a:b "subterranean dwelling" by Fi8La:L, etc.

1.3.2 Roots and Patterns: Combinatory Features

As was pointed out at the end of 1.3 above, it is characteristic of Arabic words to group into families of formally as well as semantically related forms. Another prominent characteristic to be mentioned is that derivatives of different roots may assume one and the same lexical pattern, in which case they are all rendered capable of conveying the functional sense represented by that pattern, while at the same time maintaining their respective lexical meaning. Table 1 illustrates this.

Table 1. Roots and Patterns: Combinatory Features

		Type of Pattern				
Root	Basic Meaning	Active Part. Fa:8iL	Passive Part. maF8u:L	Noun of Instrument miF8al	Locative maF8il	Verbal Noun Fa8L
ḍ-r-b	to hit	ḍa:rib	maḍru:b	miḍrab	maḍrib	ḍarb
q-b-ḍ	to seize	qa:biḍ	maqbu:ḍ	miqbaḍ	maqbiḍ	qabḍ
k-s-r	to break	ka:sir	maksu:r	miksar	maksir	kasr
f-ṣ-l	to separate	fa:ṣil	mafṣu:l	mifṣal	mafṣil	faṣl
n-s-j	to weave	na:sij	mansu:j	minsaj	mansij	nasj
n-q-b	to pierce	na:qib	manqu:b	minqab	manqib	naqb
ḥ-b-s	to hold back	ḥa:bis	maḥbu:s	miḥbas	maḥbis	ḥabs

1.3.3 Verbal Patterns

It is well known that Arabic roots are for the most part triliteral; quadriliterals, though not uncommon, are far less numerous. As reference to verbal roots is frequently needed in this study, it is necessary to mention the types of morphological patterns assumed by Arabic verbs. The following is a list of these patterns:

i) The triliteral verb and its derived patterns:

I	Fa8aL-	VI	taFa:8aL-
II	Fa88aL-	VII	?inFa8aL-
III	Fa:8aL-	VIII	?iFta8aL-
IV	?aF8aL-	IX	?iF8aLL-
V	taFa88aL-	X	?istaF8aL- (11)

ii) The quadriliteral verb and its derived patterns:

I	Fa8LaL-	III	?iF8anLaL-
II	taFa8LaL-	IV	?iF8aLaLL-

1.4 Arabic Syllabic Structure

It has often been said, especially by those concerned with foreign-language teaching and learning, that the greatest obstacle to learning a language is speaking one already. This statement, the significance of which is soon to be revealed, is relevant to the present topic in that it suggests (among other things) that the difficulties encountered in this respect are largely due to mother-tongue interference. This in turn implies another fact, namely that languages differ from one another with regard to various aspects of their structure. One of these aspects involves the distribution of phonemes or the possibilities of their sequential occurrence into the unitary groups of sounds commonly known by the name "syllables". Linguistic research has revealed that "languages differ very much in the types of syllable structures they exhibit and the places they occupy in words".(12) And of course O'Connor is right in stating – when discussing the term "phonotactics" – that "If in languages any phoneme could follow any other phoneme with equal probability there would be no place for phonotactics, but in fact there are powerful constraints operating in all languages, and each language has different rules governing the sequences of phonemes which may and may not occur."(13) Thus, to take one case, while in an English syllable the vowel may be followed by one, two, three, or four consonants (e.g. tech, techs, text, texts, -k, -ks, -kst, -ksts), other languages (such as Fijian) do not admit any syllable ending with a consonant.(14)

Similar statements may also be made about Arabic when compared with other languages, since, as would be expected, Arabic like all other languages follows its own rules and has its own characteristic types of syllable structure. This fact manifests itself clearly when it comes to using loanwords by native speakers, a process whereby the syllabic structure of such words is usually modified in accordance with that of Arabic (Chapter 4). It is not our intention to compare the state of affairs in Arabic with that in other languages; we shall concern ourselves only with the characteristic features of Arabic syllabic structure.

(i) Syllables beginning with V, CC, or C are not permissible in SA; words like *qalam* "pen", *maktab* "office", *maṭṭa:ṭ* "elastic" must therefore by syllabified as in (A) rather than (B) below:

	(A)	(B)
qalam	CV-CVC	*CVC-VC
maktab	CVC-CVC	*CV-CCVC
maṭṭa:ṭ	CVC-CVC	*CV-CVC

(ii) The vowel in a SA open syllable may be either short as in *da-ra-jah* "degree" (CV-CV-CVC), or long as in *ma:-ḏa:* "what" (CV-CV).
(iii) The vowel of a closed syllable is almost always short, e.g. *fam* "mouth", *fun-duq* "hotel", *miq-baḍ* "a handle", etc., where all the syllables are of the structure CVC. In general, it is only before a break in utterance, or pause, that a syllable of the structure CVC can be realized. A syllable of the latter type is normally avoided (i.e. unless before a pause) by being realized as CVC. For example, in a junctured phrase like *ma:+al-ᶜamal* "what is to be done", the actual utterance allows of the reading *mal-ᶜamal* with the syllabic division CVC-CV-CVC. The prepausal syllable CVC is also avoided in junctured utterance by being reduced to an open syllable of the type CV, with the remaining (final) consonant being used as the initial element of a following syllable. A case in point is the second syllable of the simple stem *bi-la:d* "country" (CV-CVC), which as a complex stem meaning "our country" would have the syllabic division: *bi-la:-du-na:* (CV-CV-CV-CV). It should be mentioned however that there are a few cases where the admissability of the syllabic structure CVC in positions where it is not prepausal is made necessary by the fact that in such cases it stands in meaningful contrast with the other type of closed syllable: CVC. Compare the following pairs of examples:

CVC $C\bar{V}C$

mar-rah "time, turn" vs. *ma:r-rah* "passers-by, pedestrians"
ᶜam-mah "paternal aunt" vs. *ᶜa:m-mah* "the masses"
ḥa:-ri-sul-mar-ma: "goalkeeper" vs. *ḥa:-ri-su:l-mar-ma:* "goalkeepers"

(iv) From the point of view of the number of syllables constituting an Arabic word, the latter can be said to consist of one of the following:

(a) One syllable, e.g. *dam* "blood" CVC
(b) Two syllables, e.g. *ġula:m* "lad" CV CVC
(c) Three syllables, e.g. *mustašfa:* "hospital" CVC CVC CV
(d) Four syllables, e.g. *ʔiktiša:fa:t* "discoveries" CVC CV CV CVC
(e) Five syllables, e.g. *mutamarridu:n* "rebels" CV CV CVC CV CVC
(f) Six syllables, e.g. *yatasa:baqu:na* "they compete with one another" CV CV CV CV CV CV
(g) Seven syllables, e.g. *ʔistiqṭa:biyyatuha:* "its polarizability" CVC CVC CV CVC CV CV CV

(v) Consonant Clusters: A consonant cluster is defined as a group of two or more different consonants not separated by any vowel. With this definition in mind, the following remarks can be made about the clustering of Arabic consonants:

(a) Clusters consisting of more than two different consonants do not occur anywhere in the Arabic (native) word.
(b) No consonant clusters are found at the beginning of words: no word begins with a syllable initiated by more than one consonant.
(c) Final clusters are possible, but only before a pause, e.g. *ᶜadl* "justice", *salm* "peace", *fikr* "thought", etc.
(d) Two-consonant clusters are common within the word, and between two vowels, e.g. *manzil* "stopping place, house", *minqa:r* "beak", *tafriqah* "discrimination".

1.5 Stress

The term stress is used for the relatively greater force or loudness exerted in the articulation of one syllable over others in the same word.

Languages vary with regard to the types of stress patterns they exhibit. Some, such as English and Russian, have Free Stress,(15) while others, like Czech, Polish and Swahili, have Fixed Stress.(16) In French, on the other hand, stress is not seen as a property of the word at all since "all the syllables in a French word are about equally stressed" and "the only increase in stress that occurs in French comes on the last syllable of the phrase".(17) Arabic may be said to belong to the second of the three language types mentioned above. Stress in it has no phonemic value, and due to its high predictability no two words are solely distinguishable by different stress placements. The following summarizes the predictability of Arabic stress on the basis of the type of consonant-vowel structure the given word has:

(a) The ultimate syllable of the word takes the stress if that syllable is of the type CVC, CVCC or CVC. Examples:

tim-'ṯa:l	statue
hi-'zabr	lion
mus-ta-'qarr	residence

Where the ultimate syllable is not one of the preceding types, stress placement is determined as follows:

(b) The penultimate syllable takes the stress provided that it is either CV or CVC. Examples:

'qa:-la	he said
ba-ra:-'hi:-nun	arguments; signs
ka:-'bu:-sun	nightmare
mi-'qaṣ-ṣun	a pair of scissors
bil-law-rah	crystal

(c) If the penultimate is of the type CV, the antepenultimate takes the stress, provided that the word contains no more than three syllables, or in cases where the number of syllables exceeds three, that the antepenultimate is of the structure CV or CVC. Examples:

'da-ra-sa	he studied
'sa-ra-qat	she stole (something)
ta-'ka:-ta-ba:	they (two) wrote to each other
ša-'kar-tu-hum	I thanked them

Stress placement in Arabic may thus be determined on the basis of the type of phonological structure the given word has. Some morphological rules, however, may also be employed for the same purpose. For instance, the first syllable is never stressed if it is the article *al-* "the" (or one of its allomorphs), or when it is one of a group of prefixes of a CV type or phonological structure. Examples:

al-'qa-mar	the moon
aš-'šams	the sun
aj-ja-'ma:l	the beauty
sa-nu-'sa:-fir	we shall travel (*sa-*, verbal prefix of futurity)
bi-ya-'day-hi	with his (two) hands (*bi-*, inseparable preposition indicating instrumentality)
fa-'qul-tu	and so I said (*fa-*, inseparable conjunction meaning "and so") (18)

Notes to Chapter 1

1 The reader may refer to Erwin (1963) where there is valuable information concerning the subject.

2 As quoted by Robins (1975), p.184.

3 Bloomfield (1973), p.178.

4 Except in contexts where we are deliberately talking about words, e.g. "What is the first word in the phrase 'the pen'?", the answer being "the".

5 It may be safely argued that the concept of word, like some other linguistic concepts, cannot possibly be defined in such a way as to apply to all languages. Rather it tends to be

defined independently in accordance with the characteristic structure of the language in question. For instance, while English "with" and Arabic "*bi-*" are both prepositions indicating more or less the same concept (as in Eng. "I wrote with the pen" and Ar. *katabtu bi-l-qalam*), unlike the former, the latter is always written (and pronounced) as part of the following element.

6 See, for instance, Ḥassān (1974), pp.224-6, and Abu al-Faraj (1961), pp.9-12.

7 In the terminology of some linguists, a word like *miḥra:r* would be described by the term "primary derivative", which is applied to any form where "none of its constituents is itself a stem, but one is a derivational affix [in this case *mi-*] and the other a derivationally bound form termed as base [in this case *-ḥra:r*]". On the other hand, a form like *ru:ḥiyy*, where one of the immediate constituents is a stem, viz. *ru:ḥ*, and the other a derivational affix, *-iyy*, would according to such linguists be termed a "secondary derivative". See for instance Hall (1964), p.176.

8 Other terms used by Arab scholars for the same concept include *qa:lab* "mould", *ṣi:ġah* "form" and *wazn* "measure". The latter, which is comparatively more commonly used than the others, basically means "weight"; to say that X is *wazn* Y is to indicate that they are of the same weight. In linguistic terminology the same term has been applied in a technical sense to indicate the morphological similarity of forms, i.e. forms assuming the same pattern. For example, words like *masbak* "foundry", *maṣnaᶜ* "factory", *maktab* "office" etc. are all nouns of place corresponding to one and the same pattern which may be symbolized as maF8al. (For the symbolization of patterns, see 1.3.1.)

9 Arab grammarians have identified a group of letters, ten in number, one or more of which may recur as additions to radical roots, thus producing new derivatives with various types of modification in form and meaning. These letters, which are traditionally known as *ḥuru:fu z-ziya:dah* "letters of addition", include *s, ?, l, t, m, w, n, y, h, a:*. These elements (or combinations of them) may be said to function as derivational and inflectional affixes as will be clearly illustrated in Chapter 2. For additional information concerning these letters, see al-Sayyid (1976), pp.31-61; Ḥasan (1963), pp.562-8; Sterling (1904), pp.13-15; and al-Afghāniyy (1969), pp.113-15.

10 Apart from loanforms, there are a relatively limited number of native items which are not analysed as containing a root and a pattern. Although they may have prefixes or suffixes added to them, these are all invariable sequences of consonants and vowels. Such items, which are to be found among the particles, include pronouns (e.g. *naḥnu* "we"), prepositions (e.g. *taḥta* "under") and conjunctions (e.g. *li?an* "in order that"). See Ḥasan (1963), p.562. For a detailed account of the meanings and uses of these particles, see Wright (1967), pp.278-96.

In his analysis of Iraqi Arabic, Erwin (1963), p.34 fn.1, refers to the elements at issue by the useful term "Solid Stems', a category in which he also includes non-native items of the type just exemplified (Erwin, p.54).

11 Arab grammarians mention a number of other forms which however to not belong to the repertory of commonly used ones in SA, and which have therefore been excluded from this list. These are forms like ?iF8a:LL-, ?iF8aw8aL-, ?iF8awwaL-, ?iF8anLaL , etc.

12 Robins (1975), p.131.

13 O'Connor (1974), p.229.

14 Robins (1975), p.132.

15 A Free Stress is one which may fall on any syllable of a word according to the syntactic function in which it is used (e.g. Eng. "export" (n.) /'ekspɔ:t/ vs. "export" (v.) /eks'pɔ:t/).

16 In languages with Fixed Stress, the position of the stress is always fixed; it always falls on a specified syllable of the word regardless of its syntactic function or other flectional changes. In Swahili, for instance, the stress is usually on the penultimate syllable.

17 That explains why French is described as a "Syllable Timed" language, the implication being that syllables, rather than only stressed syllables, tend to recur at regular intervals. See Ladefoged (1975), p.222 and Hall (1964), pp.108f. For an account of stress in Arabic, see Mitchell (1960).

18 For an exhaustive list of Arabic inseparable particles, see Wright (1967), pp.279-80, 282-3, 290-1.

2 Lexical growth through derivation and analogy

Introduction

It goes without saying that *?ištiqa:q* "derivation" in Arabic has been and still is the most important principle of word creation. It represents the most natural method by which the language has been able to generate the overwhelming majority of its native vocabulary, as well as incorporate and fully assimilate a sizeable number of borrowed elements (Chapter 4). A cursory review of the developmental history of Arabic would soon reveal the vital contribution of this tool in the process of enriching the language and enabling it to cope with the needs and demands that have faced it over the various stages of its growth. And nowadays, with the new circumstances facing it, the huge influxes of new concepts, the increasing rapidity of scientific progress, the wider range of discoveries and inventions, etc., the need to make full use of this principle is unquestionably greater than ever before.

It is an established fact that language is an adaptable sociological organism. That is to say, it undergoes processes of growth and development in the same way as does the nation or society where it is spoken. The judgements we sometimes hear that language X is primitive or language Y is advanced are often intended to imply that in both cases the degree of the evolution the language concerned has witnessed is proportional to that undergone by its environment. It remains to be said that languages may somehow vary in the way they adapt themselves, a situation which is often determined by a variety of factors such as the potential resources of the language, its place and role in the various aspects of the culture it represents, the attitude of the speakers towards their language as a cultural heritage, etc. For example: the fact that Arabic has continued to live for fifteen centuries despite all adverse circumstances is mainly attributable to religious and social factors; while, on the other hand, it is due to the nature and merits of the language itself (as a system) that it has succeeded in preserving its essential characteristics and continued to be capable of growth and expansion.

One of the aspects of growth in language is the generation of new vocabulary, which is usually the outcome of the birth or introduction of new modes of thought or the arising of new needs in society. A language may increase its vocabulary from its own essence, by depending on its native resources, or it may borrow words from other languages, either as gap-fillers or for a variety of reasons discussed in Chapter 4. It is our intention in this chapter to investigate the attempts Arabic has made over its history to enrich itself with new scientific terminology by utilizing its own means, especially the principle of *?ištiqa:q* from native roots. An important incentive behind this investigation is the hope that the study of the results obtained in the past and the practices that were then followed may assist the efforts made in our present time. We also hope to assess the role derivational processes can play under the present circumstances. Before we set about this task, however, we need to discuss a number of concepts and definitions involved.

2.1 Concepts and definitions

One of the important discoveries made by early Arab philologists, as a result of their meticulous investigation of their native vocabulary, consists in the fact that the great majority of Arabic words exhibit a tendency to group into sets or families of semantically as well as formally related forms. These relationships are more specifically defined in terms of the following observations:

(a) Apart from a limited number of items (see footnote 10 to Chapter 1), Arabic words are predominantly traceable to sequences of three (and less often four) consonant phonemes in determined order, the so-called *judu:r* "roots".
(b) The root morpheme represents one of two morphological strata out of whose combination the Arabic word may be said to result as a full lexical unit. The second stratum is represented by a sequence of vowel (and sometimes consonant) phonemes which, again, are placed in determined positions relative to the root consonants. This second stratum or component is what European writers commonly refer to as the "pattern" of the word.
(c) The morphological variations exhibited by Arabic words are in fact variations of the patterns they assume. For words of different patterns may derive from one and the same root, and where the order of root consonants is maintained all the way through (see Table 1). One of the major characteristics of Arabic relevant to this point is that it is possible to give all the conjugations of a morphological pattern without the need to use any specific root. Let us take as a case in point the second (derived) form of the triliteral verb, Fa88al-, which normally implies intensiveness (as in *qattal-* "to massacre", as opposed to the first or groundform *qatal-* "to kill", which is merely declarative):

?ana:	I	Fa88aL-tu	?u-Fa88iL-u	
naḥnu	we	Fa88aL-na:	nu-Fa88iL-u	
?anta	you (m.sg.)	Fa88aL-ta	tu-Fa88iL-u	Fa88iL
?anti	you (f.sg.)	Fa88aL-ti	tu-Fa88iL-i:n	Fa88iL-i:
?antuma:	you (du.)	Fa88aL-tuma:	tu-Fa88iL-a:n	Fa88iL-u:
?antum	you (m.pl.)	Fa88aL-tum	tu-Fa88iL-u:n	Fa88iL-u:
?antunna	you (f.pl.)	Fa88aL-tunna	tu-Fa88iL-na	Fa88iL-na

huwa	he	Fa88aL-a	ya-Fa88iL-u
hiya	she	Fa88aL-at	tu-Fa88il-u
huma:	they (m.du.)	Fa88aL-a:	yu-Fa88iL-a:n
huma:	they (f.du.)	Fa88aL-ata:	tu-Fa88iL-a:n
hum	they (m.)	Fa88aL-u:	yu-Fa88iL-u:n
hunna	they (f.)	Fa88aL-na	yu-Fa88iL-na

The simple conjugations given above represent the original example of the type of derivation generally referred to by Arab philologists as *al-ʔištiqa:qu ṣ-ṣaġi:r* "the minor or simple derivation". In view of the facts contained in the foregoing statements, Arab writers have defined this process as "the derivation of one word from another, with the two words showing the same order of root consonants and corresponding to the same general idea or meaning.(1) Words so produced may be exemplified by any set of derivatives yielded by a native single root. d-r-s, for instance, gives us *daras-* "to study", *darras* "to teach", *tada:ras-* "to study (something) together", *da:ris*"student", *mudarris* "teacher", *dira:sah* "study (n.)", *tadri:s*"teaching", *dira:siyy* "scholastic; tuitional", *madrasah* "school", *dars* "lesson", etc.

In addition to the type of derivation we have just discussed, Arab philologists have also distinguished two more types, namely *al-qalb* "metathesis" and *al-ʔibda:l* "substitution".(2) The former is based on the assumption that there exists a close connection between sounds and meanings and takes as words of the same or approximate meaning triliteral combinations of sounds, regardless of the order in which they appear. To quote one of the examples given by Ibn Jinniyy (died 1002), who is considered the most zealous exponent of this theory, the root k-l-m and the various types of combinations it yields: k-m-l, l-k-m, m-k-l, etc. are all claimed to be indicative of the same meaning, "strength and vehemence".(3) It is thus supposed that by shifting the constituent radicals around we may get a number of derivatives with more or less similar signification.(4)

With regard to *al-ʔibda:l*, the underlying assumption here is that words with two identical radicals are related in meaning despite the fact that they differ in the third radical, which may occur initially, medially or finally. The following words serve as a case in point, where the preserved radicals are /r/ and /m/:

ṣaram-	to cut off; pluck
xaram-	to pierce
šaram-	to split

etc., where the general idea is one of cutting.

ratam-	to squash, smash
raṯam-	to break the nose
rajam-	to pelt with stones

etc., where the general idea is that of breaking or crushing.

Similarly, /n/ and /b/ as the first and second radicals, give us:

nabat- (of plants) to sprout, grow
nabaᶜ- (of water) to gush forth from the spring
nabaġ- come to light, get known

etc., where the general idea is that of emergence and becoming visible.

Cases where groups of triliteral words have two radicals in common, such as the ones just cited, have been variously accounted for by different writers. The explanations put forward by those writers seems to coincide with their general views concerning the origin and growth of Arabic vocabulary in general. Al-Karmaliyy (1938, pp.1-7), for instance, maintains that Arabic words were originally biliteral forms based on the imitation of natural sounds (*muḥa:ka:t li-ʔaṣwa:ti ṭ-ṭabi:ᶜah*). These forms, according to him, through historical processes of development, have at later stages transformed into triliteral formations by having a third sound added to them initially, medially or finally. The type of sound added, he explains, has been subject to such circumstances as the nature of the particular dialect in which the development takes place, the type of idea with which the form in question becomes associated, or the type of response the concerned speakers have towards natural phenomena (al-Karmaliyy, 1938, pp.1-7).

Others, on the other hand, suggest that Arabic words are triliteral in origin, but due to processes of phonetic change which may have occurred over long periods of time and across a large number of dialects with varying pronunciations, one or more of the constituent radicals seem to have been subjected to certain alterations. Wāfi, who seems to favour this view, cites as a case in point the word *kašaṭ-* "to scratch off" as belonging to the tribe of Quraysh, but which in the dialect of Tamīm and Asad was pronounced with /q/ rather than /k/, i.e. *qašaṭ-*.(5)(6)

Wāfi's argument seems to be particularly plausible in cases where the alternating phonemes are homorganic, e.g. /ṣ,s,z/ or /t,ṯ,ḏ/. There are however instances where the alternating phonemes are articulated at different parts of the speech mechanism. These seem to be more adequately accounted for in terms of the former argument. Evidence in favour of the preceding statement may be derived from the fact that there are cases where a biliteral form still exists alongside the triliteral ones relating to it. There is, for example, *qaṭṭ-* "to cut, clip" (where the second radical has been doubled) to which there relate the forms *qaṭaf-* "to pick (flowers, fruit), *qaṭaᶜ-* "to cut off", *qaṭam-* "to break off", etc.

The above remarks concerning *al-qalb* and *al-ʔibda:l* have been made only by way of introducing the reader into the various types of word-formational processes which Arab writers have treated as belonging to the realm of *ʔištiqa:q*. However, our discussion will be for the most part confined to the type of derivation mentioned earlier, namely *al-ʔištiqa:qu ṣ-ṣaġi:r*, since, unlike the other two, the latter has continued to be the most characteristic method of word creation in Arabic and the source that represents its main potential for expansion. Therefore, the term *ʔištiqa:q* or Derivation will henceforth be used to refer to this particular type, unless otherwise stated.

The concept of /al-Qiya:s/ "Analogy"

As far as morphology is concerned, the term *qiya:s* "analogy" may be defined in simple terms as the method by which new words are formed or derived in accordance with already existing word patterns.(7) As is implied in this definition, *qiya:s* represents the set of rules which Arab philologists have inductively derived from their investigation of the native vocabulary, and according to which speakers may create new words from native roots.

The significance of the principle of analogy in Arabic derives from the fact that no sooner had the science of Arabic grammar been born than it (analogy) emerged as "a binding rule powerful enough not only to explain, but also to correct and to form. Since then, analogy ... has played a major role in the configuration of the Arabic language" (Stetkevych, 1970, p.3). It is due to the merits of this principle together with those of *ʔis̆tiqa:q* that Arabic is often likened to a mathematical formula.(8)

It is reported that the idea that language should be dominated by *qiya:s* was originally put forward early in Islamic history by a school of thought known as "al-Muᶜtazilah" (Fahmi, 1961, p.331), which was famous for its rational inquisitiveness and responsible for introducing speculative dogmatism in Islam. Arabic, according to this school, was the outcome of a convention among mortals, and had therefore to be subjected to strict scientific analysis, whereby rules and analogies were specified, authorized and observed; while exceptions were regarded as irregularities inconsistent with the agreed principles of logic.

When discussing the concept of *qiya:s* in Arabic, another concept closely related to it is inevitably given rise to. The term disignating this other concept is *as-sama:ᶜ* "(lit.) hearing", and is commonly used in opposition to *al-qiya:s*. To say that a word belongs to the category of *sama:ᶜ* is simply to imply that it has an irregular form, that it deviates from the recognized patterns or principles of derivation and that it is sanctioned by arbitrary usage of the language only, and hence represents an individual case which should not be copied.

We may mention in passing the fact that analogy in its broader sense, though a major characteristic of Arabic, is by no means peculiar to it. Languages in general resort to it as a means whereby they create new words and so enrich their vocabulary. In English, for instance, the absence of a verb to convey the verbal aspect of the concept implicit in such words as "enthusiasm" and "enthusiastic" has led to the creation of the verb "enthuse" by way of avoiding the longer expression "to become enthusiastic" or "to make enthusiastic". Again in Vulgar English we find "elocute" and "evolute" after the nouns "elocution" and "evolution" on the analogy of, for instance, execute/execution, constitute/constitution.(9)

The following discussion provides a brief account of the history behind the rise and development of Arabic vocabulary in the stages preceding the thirteenth century, which is considered as marking the beginning of its period of decline, till its second renaissance in the nineteenth century (an account of which has been given in the Introduction to this book).

2.2 Growth of Arabic vocabulary: historical background

When one studies the history of the development of Arabic and considers the rapidity of its expansion, its resistability to change (i.e. such change as has, for instance, been undergone by English over its history), and yet at the same time its qualitative capacity to adapt, and particularly its evolution from a language mainly of poetry (prior to the advent of Islam) into a successful medium of scientific expression (in post-Islamic society), one is bound to state that Arabic in this respect may well represent a unique case among the languages of the world. The state of affairs has been such that one writer (Chejne, 1969, p.52) described this rise of Arabic as one of the miracles of history. Another has referred to it as "the true wonder of Arab expansion" (Lewis, 1958, p.132). However, what interests us at this stage is to point out the fact that the changes in the religious, social and cultural life of the hitherto mainly nomadic society constituted a crucial test of the powers and potential of the language, a test which Arabic clearly proved to be highly qualified to stand. Thus, with the advent of Islam, the language which had been primarily poetic and whose vocabulary had more or less been confined to the relatively narrow world of the nomadic society,(10) was now confronted with the huge task of accommodating itself and adapting its vocabulary to an increasingly large number of new concepts and ideas pertaining to a wide variety of social and intellectual disciplines. That Arabic was quite prepared to achieve that task is a merit with which it has already been credited by many scholars of our age. Browne (1921, pp.29-31), for instance, remarks: "The old Arabs were an acute and observant people, and for all natural objects which fell under their notice they had appropriate and finely differentiated words."

It goes without saying that the Qur?ān, with the new concepts it gave birth to, the new expressions it introduced, the new values it established, was the most important factor behind the development of Arabic as well as the newly born society that was speaking it. As far as the language is concerned, we thus find that many already existing words (or at least roots) were now assigned new technical senses. For example, *kafar-*, which had previously been used to mean "to cover, conceal" was now given the meaning "to deny the existence of God, not to believe"; similarly, *fisq*, whose original signification was the "coming out of the shell" (of a ripe date), was now used with the sense of "sinfulness, dissolute life", the analogy being that a *fa:siq* "a sinful person" is one who has swerved from the commandments of God, or strayed from His course.(11)

Well before the end of the first Islamic century, the Arab armies had already expanded the area of their empire to include the major centres of civilization and become the greatest power of the time, under the second Caliph, ᶜUmar (634-644). This of course meant new challenges to Arabic, the language of the conquerors, as it had to compete with the languages of the conquered lands which were well established and backed by cultural heritage, such as Greek in Egypt and Syria and Persian in the eastern provinces. New expressions and technical terms were given rise to by the new circumstances facing the language in the fields of administration, legislation, politics, etc. In the process of creating such terms the general tendency during this stage seems to have been to resort as far as possible to native-based methods, especially derivation, semantic extension of existing words and reviving pre-Islamic vocabulary, though instances of foreign borrowings are not difficult to find (see Chapter 4). Certain types of religious expressions were also coined in accordance with the so-called *naḥt* principle (see Chapter 3).

However, it was in the two subsequent periods under the Umayyads in Syria (661-750) and the Abbasids in Baghdad (750-1258) that Arabic witnessed the most favourable factors that led to its development into a highly adequate instrument, capable of clear, simple and precise expression, besides being very rich in vocabulary, not only in the fields of Islamic studies, but also in a host of other sciences and branches of knowledge.

In addition to these developments, it is also possible to distinguish a number of other significant factors involved in the growth of Arabic vocabulary. These include:

(i) The Arabs' love for their language and their enthusiasm to excel in studies related to it continued unabated despite the radical changes that had taken place in their life and environment. Thus as early as the eighth and ninth centuries there appeared two philological schools at Basrah and Kufah (in Iraq), both of which were later united with the school of Baghdad.(12)

(ii) As an outcome of the significant developments in the religious, political and cultural life of the new society there now existed a number of theological schools such as the aforementioned "al-Muᶜtazilah", and politico-religious movements such as "al-Khawārij".(13) The issues these and other movements concerned themselves with revolved around questions like fatalism or predestination *aj-jabriyyah* as opposed to freedom of will *al-qadariyyah*, God and his nature, the Muslim and his status after he has sinned, the Muslim state and who should head it, and so on. The contribution of these movements to the enrichment of Arabic consists in the new expressions and technical terms that must have been introduced by way of expressing, delimiting or defining the various views and stands advocated by these schools.

(iii) The Abbasid period can be said to have witnessed the most favourable circumstances for the growth of scientific and technical terminology in Arabic in its early history. One of the most outstanding features of this period was the outburst of intellectual activity such as the east had never witnessed before. We can do no better to illustrate the situation than to quote Nicholson (1953, p.281), who has this to say:

> It seemed as if all the world from the Caliph down to the humblest citizen suddenly became students, or at least patrons of literature. In quest of knowledge men travelled over three continents and returned home, like bees laden with honey, to impart the precious stores which they had accumulated to crowds of eager disciples, and to compile with incredible industry those works of encyclopaedic range and erudition from which modern science, in the widest sense of the word, has derived far more than is generally supposed.

Thus with this great enthusiasm for learning, the impact of these circumstances on the language cannot be underestimated. The language now was highly successful not only in terms of its capability to provide adequate terminology in Arabic or native sciences, but also in terms of its being in possession of such qualties as clarity, precision, and the capacity for abstract expression in various fields of knowledge, including those with which the Arabs were acquainted through cultural contact with other nations.

As their fields of interest began to grow both in number and variety, Arab scholars, as far back as the tenth century, must have found it useful to distinguish between two main categories of science. There are thus (a) the Native or Arabic Sciences, which include the traditional or religious ones (*al-ᶜulu:mu l-naqliyyati ?awi š-šarᶜiyyah*) and (b) the Foreign Sciences, in which they included the intellectual or philosophical sciences (*al-ᶜulu:mu l-ᶜaqliyyati ?awi l-ḥikmiyyah*).(14) This general categorization may be elaborated by the following table, which nevertheless does not claim exhaustiveness:(15)

Native Sciences	
al-fiqh	jurisprudence
at-tafsi:r	Qur?anic exegesis
al-ḥadi:ṯ	prophetic tradition
al-kala:m	theology
al-baya:n	rhetoric
an-naḥw	grammar
aṣ-ṣarf	morphology
al-?adab	belles-lettres
al-luġah	lexicography
al-qiya:s	analogy
Foreign Sciences	
al-falsafah	philosophy
aṭ-ṭibb	medicine
al-handasah	geometry
al-falak	astronomy
al-mu:si:qa:	music
al-ki:mya:?	chemistry
ar-riya:ḍiyya:t	mathematics
al-ḥiyal	mechanics

(iv) Far beyond the borders of its homeland, Arabic was also enjoying a very prosperous period. In Spain, for instance, no sooner had the Arab conquerors settled than they established their mother tongue as the official language in which documents, sermons, coins, etc. were issued. The soil was also equally fertile for Islamic civilization and culture, with which thousands of Christian Spaniards became rapidly imbued, so much so that one of the bishops of Cordova could not help raising this bitter complaint (quoted by Dozy, 1913, p.268):

> My fellow-Christians delight in the poems and romances of the Arabs; they study the works of Mohammadan theologians and philosophers, not in order to refute them, but to

> acquire a correct and elegant Arabic style. Where today can a layman be found who reads the Latin commentaries on Holy Scripture? Who is there that studies the Gospels, the Prophets, the Apostles? Alas! the young Christians who are most conspicuous for their talents have no knowledge of any literature or language save the Arabic; they read and study with avidity Arabian books; they amass whole libraries of them at a vast cost, and they everywhere sing the praises of Arabian lore. On the other hand, at the mention of Christian books they disdainfully protest that such works are unworthy of their notice. The pity of it! Christians have forgotten their own tongue, and scarce one in a thousand can be found able to compose in fair Latin a letter to a friend! But when it comes to writing Arabic, how many there are who can express themselves in that language with the greatest elegance, and even compose verses which surpass in formal correctness those of the Arabs themselves!

The reader will now have realized that the process of growth and expansion experienced by Arabic in its early stages was enhanced, besides its internal qualitative capacities, by what may be described as favourable external factors and circumstances such as its being the language of the Holy Qur?ān and the widely spreading religion, the extensive social, cultural and particularly scientific contact between the Arabs and other nations, the great interest taken by early scholars in studies involving its various aspects, etc. Further aspects of the cultural contact between the Arabs and other nations and the impact of this on language will be discussed in Chapter 4.

However, the circumstances Arabic has undergone have not always been favourable. It could be said in general (and has already been pointed out) that from the end of the thirteenth until the nineteenth century Arabic experienced a state of semi-hibernation, which could be accounted for as being a natural result of the intellectual stagnation characteristic of life and people in that period. That state, in turn, was due to the fact that the Arabs were now recurrently invaded, weakened and eventually dominated by different forces: Mongols, Turks, Berbers, etc. In 1258, for instance, Baghdad, known as *madi:natu s-sala:m* "City of Peace", with a well established reputation for being a centre of intellectual activity and civilization, was swarmed over and violently ravaged by Mongol hordes under Hulagu.

Evidence of the fact that Arabic during that period was witnessing what may be described as its Dark Ages can be derived, among other things, from a situation similar to that mentioned earlier concerning the complaint raised by the bishop of Cordova, except that this time the picture is reversed and the one complaining is an Arab. The fourteenth-century traveller, Ibn Baṭṭūṭah, on his visit to al-Baṣrah in 1327, and having listened to a preacher giving a sermon, describes in the following words his impression about that preacher's command of Arabic (quoted by Gibb, 1963, p.142):

> I attended once the Friday prayers at the Mosque, and when the preacher rose to deliver his sermon, he committed many serious grammatical errors. I was astonished at this and spoke of it to the qāḍī [i.e. the judge of that city], who answered, "In this town there is not one left who knows anything about grammar". Here indeed is a warning for all men to reflect on — Magnified be He who changes all things and overturns all human affairs! This Basrah, in whose people the mastery of grammar reached its height, whence it had its origin and where it developed, which was the home of its leader [Sībawayh, the leader of the Basrah school of grammar] whose preeminence is undisputed, has no preacher who can deliver a sermon without breaking its rules.

Following this period of darkness, in the nineteenth century and as if suddenly awakened to a bitter reality, the Arabic-speaking world started a new age of revival and flourishing, which has confronted Arabic once again with a situation similar to that of its earlier renaissance. The historical background to this new renaissance has already been discussed in the Introduction to this book. Meanwhile, having formed an idea about the historical factors involved in the growth of Arabic, let us turn to the language itself and the various aspects of adaptation and development it has witnessed due to the impact of these factors and the ever-growing need for new terminology. In the following discussion we shall examine the process of lexical adaptation in Arabic in the first renaissance.

2.3 Lexical adaptation in the first renaissance

It has been suggested above that Arabic has the capacity to adapt itself to the needs and circumstances facing it. The present investigation seeks to illustrate the preceding statement with examples of scientific and technical words used in the period following the advent of Islam until the thirteenth century. Our examples are taken from various subjects and relate to both the categories of science (native and foreign) mentioned above.

As far as the process of creating new vocabulary is concerned, the distinction between native and non-native sciences may be said to derive some significance from the fact that, generally speaking, in the case of native sciences the process in question has not constituted a real challenge to Arabic. For, in order to furnish new terms for the new concepts introduced by these sciences, it has followed its natural course, applying its characteristic method of word formation and utilizing its native roots as sources of derivation. In the case of foreign sciences, on the other hand, the situation cannot be said to have been the same. That Arabic came into contact with these sciences mainly through translation, as we shall see in Chapter 4, is indicative of the fact that it was somehow subjected to the influence of a foreign language or languages, at least as far as the lexicon is concerned. And despite the hard efforts to keep Arabic pure, traces of this influence are still alive; suffice it to say that the names Arabic uses for some of these sciences are themselves non-native, e.g. *falsafah*, *handasah*, *musi:qa:*, etc. This is the case despite the fact that in many instances the traces have been gradually wiped out. For instance, the sense in which in the time of al-Rāziyy (d.1209) was denoted by the Greek loanform *hayu:la:* (see al-Rāziyy, 1343 AH, vol.2, p.41), is now currently conveyed by the native *ma:ddah* "matter". Further examples of this sort are provided in Chapters 3 and 4.

However, the statement that traces of non-native influence are detectable in Arabic is in no way intended to imply that it was unsuccessful in meeting its needs for scientific vocabulary, not only in the indigenous sciences but in the assimilated ones, too. An examination of the tenth-century scientific vocabulary would reveal the fact that Arabic has not failed to create its own terminology. In chemical vocabulary, for instance, we find that various types of chemical processes are clearly expressed through the application of native patterns. One of the most recurrent of these patterns is taF8i:L from the verbal Fa88aL-. Here are some instances of the early occurrence of this pattern:

taqṭi:r	distillation	q-ṭ-r
taṣᶜi:d	sublimation	ṣ-ᶜ-d
taḥli:l	solution	ḥ-l-l
tašmi:ᶜ	ceration	š-m-ᶜ
taṣwi:l(16)	lixiviation	ṣ-w-l
tarxi:m(17)	incubation	r-x-m
taxni:q(18)	constriction	x-n-q

This capability of Arabic morphological patterns to meet the needs of scientific language is an indication of the fact that they possess those significatory featureswhich are indispensably required in scientific terminology. The pattern we have just exemplified, for instance, besides its implication of such ideas as intensiveness and extensiveness, which are often applicable to scientific (especially chemical) processes, is also capable of causative signification as well as denominative derivation. Thus from an intransitive form like *qaṭar-* "to drip" we may derive the causative verb *qaṭṭar-* with the technical sense "to refine or distill", a corresponding verbal noun *taqṭi:r* "distillation". And the already cited *tašmi:ᶜ* "ceration" relates to *šammaᶜ-* "to apply cerate", which is a denominative from *šamᶜ* "wax".

What has been said above concerning the pattern taF8i:l and its capability of scientific expression is also true of the other patterns of Arabic. The following instances have been attested in early literature (these patterns are illustrated by single examples since other instances of words assuming these patterns are provided later):

Fa8L	*sabk*	casting or melting
Fa8aLa:n	*ġalaya:n*	boiling
?iF8a:L	*?ilġa:m*	amalgamation
?inFi8a:L	*?injima:d*	solidification
taF8iLah	*taṣdi?ah*	rusting
taFa88uL	*taᶜaqqud*	coagulation

The observations we have so far made with regard to the process of lexical adaptation in the field of chemistry are also applicable in other disciplines. Here are some examples from the medical vocabulary of the Abbasid period (see Zaydān, "al-Lughah", pp.81-82):

?istiF8a:L	*?istisqa:?*	dropsy
Fa8aLa:n	*xafaqa:n*	palpitation of the heart
taF8i:L	*tašri:ḥ*	anatomy
?iFti8a:L	*?ixtila:ṭ*	mental disorder
taFa88uL	*tašannuj*	spasm

From the terminology of physics and metaphysics we may cite the following instances as used by Fakhr al-Dīn al-Rāziyy (1343 AH, pp.1/133, 1/128, 2/397, 2/322, 3/121):

Fu8u:L	*ḥudu:ṯ*	incidence
Fa8aL	*ᶜadam*	non-being; non-existence
taFa:8uL	*tana:sux*	metempsychosis
?istiF8a:L	*?istifra:ġ*	exerting oneself to the utmost
Fu8u:Lah	*luzu:jah*	elasticity

It may be appropriate at this stage to bring into focus a point which the reader may have already noticed, namely that an important aspect of lexical adaptation exhibited by medieval vocabulary is the semantic extension or figurative use of already existing radical concepts. And here more than one derivative of the same root may be used to convey certain specialized meanings which are somehow different from one another, though still related to the same basic idea of the root. Let us take as a case in point the last of the examples given on the list in the middle of the previous page, i.e. *taxni:q* "constriction" and before starting our discussion, we may just as well consider the following context in which this word is used by an early Arab scientist:

> [Glass] flasks are required in the process of constricting [*taxni:q*] those things that can be sublimed. For alchemists may require to constrict a substance, so they place it in the flask and cause it to ascend, and consequently it ascends [in the flask] and is constricted in the neck.(19)

The original signification of the root x-n-q is "to strangle, squeeze the throat" and all its derivatives in the literary language revolve around the same concept or something close to it. For example, *xina:q* is "a cord or string with which one is strangled", *xanna:qah* "a snare with which beasts of prey are taken by the throat", etc. The use of the derivatives of this root for the chemical process illustrated in the above passage (e.g. *taxni:q* "constriction", *taxannaq-* "to be constricted") is no doubt a result of the analogy between the neck of a person or animal and that of a flask. We may also add, incidentally, that outside chemical vocabulary, other derivatives of the same root have also acquired certain specialized senses. For example, *?ixtina:q* has been applied to such medical conditions as asphyxia and the contraction of the womb muscles. More recently, the verbal forms *xanaq-* and *?ixtanaq-* have assumed a technological signification, to throttle or be throttled down (of an engine). And last but not least, in political jargon we hear nowadays of *xanqu l-ḥurriyya:t* "the suppression of liberties".

Thus through derivation, semantic extension as well as other means of neologization which we shall discuss later, the scientific and technical terms used by medieval scholars soon came to be numerous and varied — so much so that a number of philologists saw a need for devoting whole works to the compilation, classification and definition of these terms. The first attempt of this kind was successfully carried out by a writer we have already cited, al-Khuwārizmiyy (d.997), who also reveals himself to be fairly well informed on the nature and characteristics of technical terminology.(20) In the introduction to his "Mafātiḥ al-ᶜUlūm" ("Keys to the Sciences") for instance, he points out that the technical sense of a word may vary from one science to another. As a case in point, the noun *rajᶜah* r-j-ᶜ "to come back" may convey the following senses according to the type of discipline in which it is used (al-Khuwārizmiyy, 1895, p.3):

in lexicography	single return
in jurisprudence	marrying a divorced wife again
in theology	the return of the expected Imām(21)
in astronomy	the retrocession of a star

2.3.1 Features of lexical development in the first renaissance

The following are a number of significant developmental and adaptational phenomena featuring in Arabic in its first renaissance:

(i) An easily detectable feature in the Arabic of the seventh to thirteenth centuries is the wide use of a morphological device, hitherto far less widely employed, whereby it could create a large number of abstract substantives denoting non-material concepts, or the idea of things as distinguished from the concrete things themselves. The device at issue consists in the use of the nominal suffix *-iyyah*, which yields types of words corresponding to English substantives ending in -ism, -ity, -ness, etc.; for instance: *?insa:niyyah* "humanity, humaneness" from *?insa:n* "a human being".

It has been noticed that prior to the first cultural and scientific renaissance, this suffix had hardly been used for abstract expression. This may be accounted for by the fact that the simplicity of the life and activities the Arabs experienced in that period did not call for minute investigation on their part into the nature and essence of things and concepts in their environment. In the subsequent periods, however, this state of affairs changed drastically. The engagement in close reasoning and rigorous study of a wide variety of disciplines, and hence the acquisition of a more thorough and precise look at things, must have motivated a similar trend in the language. Thus we find that prior to this development the common way of expressing abstract notions had been by using long periphrastic expressions. For example, instead of saying someone had done something out of *xayriyyah* "charitableness" or *ᶜadliyyah* "impartiality", etc., the predominant way of putting it was to say *ᶜala: jihati l-xayr* "in the manner (or for the sake of charity", *ᶜala: jihati l-ᶜadl* "in the manner of justice", and so on (see MMLᶜAM, vol.i, 1935, p.212). Later stages, however, witnessed the use of the suffix *-iyyah* as a useful tool for denoting abstractions. The following examples are only a few instances of the wide application of this suffix in those later stages:

nawᶜiyyah	specificity	<*nawᶜ*	kind, species
šaxṣiyyah	individuality(22)	<*šaxṣ*	person, individual
dahriyyah	atheism(23)	<*dahr*	time, eternity(24)

The same suffix was also added to passive participles thus creating doublets like:

ᶜaqliyyah	mentality	<*ᶜaql*	intellect
maᶜqu:liyyah	intelligibility	<*maᶜqu:l*	intelligible
ᶜilliyyah	causality	<*ᶜillah*	cause
maᶜlu:liyyah	causeity	<*maᶜlu:l*	caused(25)

A significant development to be noted here is that abstract substantives were derived not only from nouns, although this was predominantly the case, but also from pronouns and particles like adverbs and prepositions which are rarely used in derivational processes in Arabic. There are, for instance:

ʔana:niyyah "egotism" < *ʔana:* "I"
huwiyyah "essence; substantiality" < *huwa* "he"
ma:hiyyah "quality; what a thing is" <*ma:* "what" + *huwa* "he/it"
kayfiyyah "abstract quality" < *kayfa* "how"(26)
laysiyyah "non-being; nothingness" < *laysa* "not to be"
maᶜiyyah "simultaneity" < *maᶜa* "with"(27)

Another similarly but less frequently used termination is *-u:t*, which Arabic seems to have acquired from Aramaic (see Wright, 1967, vol.i, p.166 and al-Khuwārizmiyy, 1895, p.34), e.g. *la:hu:t* "divinity", *na:su:t* "humanity", *raḥamu:t* "compassion", *malaku:t* "power and majesty", etc.

(ii) We have said elsewhere (Chapter 3) that it is uncharacteristic of Arabic words to be compounded with one another, as is the case, say, in English or Greek. Our investigation of medieval vocabulary, however, has shown that constructions of this nature have not been altogether non-existent, though we must hasten to say that the phenomenon in question has been extremely sporadic and restricted only to certain cases. A case in point worthy of mention here is the insertion of the negative particle *la:* "no" between the definite article *al-* and the noun it defines. There are for instance:

al-la:niha:yah	"infinity"	< *la:-* + *niha:yah* "end"	
al-la:ḍaru:rah	"non-necessity"	< *la:-* + *ḍaru:rah* "necessity"	
al-la:ʔadriyyah(28)	"scepticism"	< *la:-* + *ʔadriyyah* "knowledge"	

The occurrence of such formations, it has been rightly argued, is due to the influence of corresponding Greek constructions (Wright, 1967, vol.i, p.166; al-Khuwārizmiyy, 1895, p.34). It remains to be added that in present-day Arabic, neologisms of the type just exemplified are on the increase in a wide variety of disciplines. Here are a few instances:

al-la:nafa:ḏiyyah	impermeability	physics
al-la:talafiyyah	indestructibility	physics
al-la:?ixtira:qiyyah	impenetrability	physics
al-la:filizz	non-metal	chemistry
al-la:?inṣiha:riyyah	infusibility	chemistry
al-la:ᶜunf	non-violence	politics
al-la:salm	absence of peace	politics
al-la:ḥarb	absence of war	politics
al-la:?inṯina:?iyyah	inflexibility	maths
al-la:qiya:siyyah	incommensurability	maths
al-la:šuᶜu:r	the unconscious	psychology

(iii) As an alternative to the use of compound words, early scientists resorted to syntactic constructions or more-than-one-word lexical units. This seems to have been the case where scientists felt the need to give a precise description of what they wanted to indicate and where single-word terms capable of faithfully rendering the idea were not available. Thus, for a chemical instrument which he describes as "a stove with a base supported on three legs, the walls and bottom being perforated with holes", al-Khuwārizmiyy (pp.257-8) uses the term *na:fixu nafsih* "(lit.) that which blows itself". As far as literary Arabic is concerned, constructions of the type just illustrated are traceable to pre-Islamic times. The nickname of a famous poet of al-Jāhiliyyah,(29) Thābit bin Jābir, was *ta?abbaṭa šarran* "he carried mischief under his arm"; another is *baraqa naḥruh* "his throat shone".

(iv) A fully detailed account is given in Chapter 4 of the phenomenon of lexical borrowing in medieval as well as modern Standard Arabic with particular attention being paid to its relevance to the vocabulary of science and technology. There is one aspect of this question which relates to the present investigation and calls for some consideration. This is the practice, now fairly established in Arabic, of modifying foreign borrowings in accordance with the characteristic morphological structure of the language. The point most relevant in this connection is that by subjecting these borrowings to its system of analogical derivation, Arabic has succeeded not only in giving them an Arabic shape, but also in using them for further derivation – so much so that many such words are now hardly distinguishable from native ones. Examples are amply provided in Chapter 4; here we will look at one case in point: A loanform commonly used by medieval Arab thinkers derives from the Greek word "philosophos" (philosopher"; the Arabic rendition of this loanform is assigned the native pattern Fay8aLu:L thus producing *faylasu:f* on the analogy of such a word as *ḥayzabu:n* "old hag". Such being the case, the word *faylasu:f* is also analysable in terms of being based on the quadrilateral root f-l-s-f, which is abstracted from the original (Greek) form. Now that the root is available, the process of producing further derivatives from it becomes no less natural and automatic than that involving use of native roots. As a result of applying this process there now exist in Arabic the forms:

falsaf-	to philosophize	Fa8LaL-
tafalsaf-	to practise philosophy; to pretend to be a philosopher	taFa8LaL
falsafah	philosophy	Fa8LaLah
falsafiyy	philosophical	Fa8LaLiyy
fala:sifah	philosophers	Fa8a:LiLah
tafalsuf	philosophization	taFa8LuL
mufalsif	philosopher (act.part.)	muFa8LiL
mutafalsif	philosophaster	mutaFa8LiL
mutafalsifu:n	philosophasters	mutaFa8LiLu:n

On the other hand, loanforms that have not received the same treatment are easily recognizable as non-native, let alone their incapability of yielding further derivatives.

(v) Contrary to what is claimed by many writers and philologists, there is much linguistic evidence that concrete nouns were used by early Arabs as sources of derivation. The argument put forward by such people is that concrete nouns may not lie at the basis of Arabic derivation, i.e. be the basis of verbal derivations. We have noticed, however, that at least as far as scientific vocabulary is concerned, in a large number of cases scientists have not refrained from using these nouns to produce new derivatives wherever necessary. There are, for instance: *ḏahhab-* "to gild" < *ḏahab* "gold", *za?baq* "to rub coins with *zi?baq* 'mercury'", *muzaffat* "smeared with *zift* 'pitch'". Due to the importance of the process in question and its great impact on the growth of present-day Arabic scientific vocabulary, we will give in section 2.4 a fully detailed account of this issue and the various views involved in it.

(vi) It is widely held that Arabic is extremely rich in synonyms, that it uses a multitude of synonymous words for denoting the same object or concept. However, the question of whether these so-called synonymous words are really identical in meaning has rarely been subjected to close investigation. Such an investigation is bound to reveal the fact that there does exist a difference of meaning; it is only that this difference is sometimes extremely fine and that in many contexts the distinctions are not of importance to the subject of discourse. Tangible proof of this claim can be derived from a close examination of any group of such semantically related words. What we are driving at in this connection is to point out the fact that this wealth of near-synonyms has proved to be of great help for early scholars in that they used these words to express the finest shades of meaning and the most delicate conceptual distinctions. It is of course due to those scholars' scientific and analytical methods of investigation that what in ordinary or literary language were more or less mutually interchangeable words, came to be distinguished within the same semantic field to which they relate.

Let us take as a case in point a group of words relating to the concept of knowledge and see how early philosophers and thinkers used them to differentiate the various aspects involved in this concept.

Words expressing sensible knowledge may thus be found clustering around the root ḥ-s-s "to sense", such as *al-ḥiss*, which indicates the faculty of feeling or perceiving, i.e. "sense", *al-ʔiḥsa:s* "sensation, the action of perceiving through the senses", *al-maḥsu:s* "that which is perceptible through the senses", etc. (See Sprenger, 1854, p.81, fn.1; pp.302-8.)

There is also the faculty of knowledge through consciousness for which Arabic has the word *aš-šuᶜu:r* š-ᶜ-r. Some philosophers define this concept as the first stage at which knowledge is acquired (Sprenger, p.746).

A third faculty is that of *al-wahm* w-h-m, which involves the perception of abstract ideas and qualities pertaining to perceptible entities (*maḥsu:sa:t*). It is through this faculty that one acquires the feeling, e.g. that John is brave or generous. Some aspects of animal behaviour are also accounted for in terms of this faculty. For instance, it is through this faculty that, say, a sheep acquires the awareness that a wolf is to be escaped from (al-Jurjāniyy, 1357 AH, p.228).

at-taxayyul x-y-l is a further faculty whereby is indicated the act of preserving in the sense of the qualities or images of what has been known through sensory perception. For instance, if we see a picture for a second time, after we have completely forgotten it, it is this faculty that enables us to realize that this is the same picture we saw first (Sprenger, pp.304, 451, 453).

There is also *al-ʔidra:k* d-r-k which indicates the comprehension by the mind of the different kinds of abstraction and therefore encompasses perception through *al-ʔiḥsa:s*, *aš-šuᶜu:r*, *al-wahm* or *at-tawahhum* and *at-taxayyul*.

As to *at-tajri:d*, this is a word whereby is denoted the "purely intellectual abstraction, the comprehension of the universal concept, the purely intelligible" (Goichon, 1969, p.67).

Last but not least there is *al-fiṭrah*, which indicates man's natural disposition to understand and also refers in a general way to a person's characteristic nature or innate character.

These cases by no means cover all the Arabic words that may be said to relate to the semantic field of knowledge. However, we may as well stress the point that the adaptation of Arabic literary words to scientific and technical signification has been an increasingly growing process, especially in our own time. It is also true that Arabic in the first renaissance was successful in meeting its needs for scientific terminology through its native system of derivation and by resorting to its own wealth of words and roots.

By way of conclusion, it may be relevant to point out that an important factor in the adaptation of literary Arabic to scientific needs was the fact that many of the early Arab scientists were of considerable linguistic sophistication, well informed on the potential resources of their language and had a good command of its vocabulary. This being the case, they had the opportunity, as far as possible, not to leave the door open to non-native types of lexical structures or foreign vocabulary. Some were even distinguished linguists whose interest in the literary aspects of the language was no less notable than their devotion to their science proper. Muḥammad bin Zahar (d.596 AH), Muḥammad bin Rushd(d.595 AH) and Ibn Sīna (d.1037) are a few examples (see Ḥusayn, 1060, p.231).

2.3.2 Morphological patterning and early scientific vocabulary

Arabic has the well established reputation for forming its vocabulary according to certain morphological patterns which are themselves capable of certain types of lexical signification. This constitutes one of the merits with which Arabic is often credited. Referring to the usefulness of the pattern Fu8a:L in generating new medical vocabulary, Browne (1921, pp.35-6) has described the process as "a great power of Arabic". This method, he argues, enables you to know what a word means even though you are meeting it for the first time (provided, of course, that you know the basic idea embodied by the root). Thus, he goes on, "I never met with the word *jubal* ... from *jabal*, 'mountain', but if I did meet with it, I should know that it could mean nothing else but 'mountain sickness'."

The denotive capacity of such patterns is obviously a result of their frequent occurrence in the language with different roots, but always the same denotative function. The pattern in question for instance is assumed by a great majority of names of diseases and ailments. Here are some examples taken from early medical literature (see al-Tha^{c}ālibiyy, 1938, pp.136-40; and Appendix):

xuna:q	diphtheria	x-n-q	choke, strangle
qula:^{c}	canker of the mouth	q-l-^{c}	pull out, tear
juḏa:m	leprosy	j-ḏ-m	cut off, maim
duwa:r	giddiness	d-w-r	revolve
suba:t	lethargy	s-b-t	sleep, rest

There is also the pattern Fa8aL which is used for the same signification (see al-Khuwārizmiyy, pp.157, 160, 163; and Appendix):

salas	incontinence of urine	s l s	to run continually
ḥafar	decay of the teeth	ḥ-f-r	to dig
sabal	cataract of the eye	s-b-l	to let fall

The following are other recurrently used patterns with examples of their occurrence in early scientific literature (there are further examples in the Appendix):

Fa8aLa:n for denoting motion, commotion, fluctuation, etc.:

dawara:n	rotation
fawara:n	boiling
zayaġa:n	swerving, deviation
xafaqa:n	palpitation, fluttering
sayala:n	flowing, gonorrhea

Fu8a:Lah	for denoting small portions which are broken, thrown away:	
bura:dah		filings
bura:yah		shavings
quṣa:ṣah		parings
kusa:rah		broken pieces
quṭa:ᶜah		cuttings
ruḍa:ḍah		pounded fragments
quma:mah		sweepings

Fa8i:L	for denoting sounds:	
zafi:r		sound of exhaling
šahi:q		sound of inhaling
faḥi:ḥ		hissing (of a snake)
ḥafi:f		rustling (of dry leaves etc.)
ṣafi:r		whistling

Fi8a:Lah	for denoting activity, profession, post etc.:	
ḥija:mah		cupping
diba:ġah		tanning, tanner's trade
nija:rah		carpentry, woodworking
xiya:ṭah		sewing
giya:dah		leadership

miF8aL, miF8a:L, and miF8aLah, these are what Arab philologists designate as denoting nouns of instrument:

miṭmar	plumb-line
midwas	a tool for polishing
misṭaḥ	dough-roller (30)
mihra:s	mortar (for pounding)
minfa:x	bellows (31)
misbakah	mould
miḥjamah	cupping glass (32)

A fully detailed account of the role and significance of such patterns in the configuration and development of Arabic scientific vocabulary is provided in the following discussions, with special emphasis on the present state of affairs. Meanwhile, while recognizing the high recurrence of these patterns in the early history of the language, we have been able to detect a number of cases worthy of mention. A careful examination of early vocabulary is bound to reveal the fact that the concurrence of the morphological patterns with their respective meanings has not been a consistently observed process. A noun of instrument, a noun denoting motion, a name of a disease, etc. may be denoted by words assuming morphological patterns other than the ones mentioned above. For instance, there are a number of cases where words denoting names of diseases have patterns other than Fu8a:L or Fa8aL:

hayḍah	cholera	Fa8Lah
xilfah	diarrhoea	Fi8Lah
fatq	hernia	Fa8L
yaraqa:n	jaundice	Fa8aLa:n
?istisqa:?	dropsy	?istiF8a:L

More-than-one-word lexical units have also been used for the same purpose, e.g.:

ᶜirqu n-nasa:	sciatica
da:ʔu l-fi:l	elephantiasis
ḏa:tu r-riʔah	pneumonia
ḏa:tu j-janb	pleurisy

It may also be useful to add that some medical words were also borrowed from other languages, such as Greek, e.g. *niqris* "gout", *kaymu:s* "gastric juice", *qawlanj* "colic" (see Ḥusayn, 1960, p.238).

On the other hand, a word assuming the pattern Fu8a:L or Fa8aL need not necessarily always be denotative of a disease or ailment. Here are a few examples:

Fu8a:L		Fa8aL	
ruwa:ʔ	grace, prettiness	*karam*	generosity
suᶜa:ᶜ	ray, beam	*barad*	hail
ṣura:ḥ	pure	*qasam*	oath

There are even cases where a non-native word has been used despite the fact that there exists a native root from which an analogical form could have been derived. For instance, the Arabic word for the instrument known in English by the name "pen" has not been derived from the widely employed root k-t-b "to write", i.e. there exists no such form as *miktab*, *mikta:b* or *miktabah*, with the meaning "pen", at least not as far as I know. Rather, the instrument in question is denoted by the loanform *qalam* < Greek kalamos (see Fahmi, 1961, p.176). The native *yara:ᶜ* "reed pen" is hardly ever used nowadays, except perhaps in poetic language.

Before we go on to the next stage, and by way of introducing the reader to it, we need to draw attention to the fact that Arab language academies as well as individual writers and many people concerned with the development of modern scientific vocabulary in Arabic attach great importance to analogical derivation and the employment of morphological patterns as a means of lexical self-enrichment. This, no doubt, is quite natural and justifiable since it represents a tendency towards preserving the symmetry and homogeneity of the language. There is no denying, however, that the views and proposals put forward in this respect are sometimes far from being realistic and reflect a considerable degree of extremism and unjustifiable purism — so much so that many such attempts which are supposed to be aimed at reviving the language often turn out to be more of a hindrance than a help. The following discussion will be concerned with the evaluation of a number of the most important ideas and suggestions held or advocated by academicians and other concerned scholars.

2.4 Analogical derivation and modern scientific vocabulary

2.4.1 The role of language academies

It could be generally stated that the policy of Arab language academies concerning the growth of Arabic vocabulary has been characterized by a strong tendency towards adopting as sources of inspiration and models of imitation the linguistic achievements of medieval philologists and grammarians. Modern neologisms that do not conform to the rules and patterns established by those early scholars are often frowned upon and rarely accepted, and if so, are treated as special cases. Certain measures have, however, been taken by way of prompting and encouraging certain evolutionary features, thus aiming at stepping up the process of creating new vocabulary. It should not be understood, though, that these measures have constituted any drastic change in the early established principles of word formation. What has almost always been the case is that certain types of word-formational procedures common to Medieval or Classical Arabic (some of which have already been discussed) have now been sanctioned and recommended as being analogically applicable, and may thus be taken as models for further creations. In some cases, the same has been applied to types of formation that have not been widespread in early Arabic or which the latter has developed due to external linguistic influence. A few other developmental features common to present-day Arabic and which may be said to have occurred in imitation of non-native scientific vocabulary, have also been approved, though with certain reservations.

The fact still remains however that much controversy is involved in the treatment of various aspects of the process of dealing with lexical deficiency through the utilization of native resources. This controversy is noticed within language academies as well as outside them, among individual linguists and writers. The present discussion is a detailed and critical account of the attitudes of Arab language academies regarding a number of important issues relevant to our investigation, with special reference being made to the issues as debated within the Arabic Language Academy of Cairo.(33)

One of the important tasks to which language academies in general set themselves is to regulate the processes through which neologisms are created and (at least as far as Arab academies are concerned) to keep these processes in line with the characteristic system of the language concerned as far as possible. This has been one of the most prominent features characterizing the efforts of language academies in the Arab world. The Cairene Academy for instance has issued a number of decrees whereby it declares certain word-formational processes as analogically applicable, i.e. they may be taken as models for similar coinages.(34) Such a step has been seen as extremely essential, for there has been and still is considerable discrepancy concerning the analogicality of certain lexical patterns. By so doing, the academy has also defined its own position (as a responsible and authoritative body) regarding what may or may not be permissible in this respect. The following is our account of some issues in this controversy.

(i) Derivation from concrete nouns

It has already been hinted that many writers are opposed to the employment of concrete nouns for verbal derivations and that already existing instances of such derivations are considered non-analogical. This position is based on the argument that the number of instances in the language where a concrete noun (*ʔismu ᶜayn*) lies at the basis of derivation is relatively limited compared to cases where the basis is an abstract noun (*ʔismu maᶜna:*). It is also argued that the basis of derivation should be a noun expressing an action or state without any reference to person, time or place. This noun – technically known as *al-maṣdar* "the starting-point" – cannot but be, the argument goes, an abstract and not a concrete noun (see al-Iskandariyy, 1935, p.233).

The academy's debates on the matter resulted in issuing a decree whereby it deems derivation from concrete nouns as permissible, but only in so far as this is impelled by necessity, with the further restriction that it should be confined to the language of science only (see Appendix).

It cannot be denied that the academy's attitude towards the problem and as revealed by the position just explained involves a certain degree of unwarrantable reservation. For one thing, derivation from concrete nouns, contrary to what is frequently alleged, has not been an insignificant phenomenon in early Arabic. Instances of such derivations have already been cited, and it may be useful to point out that some morphological patterns have been distinguished for their frequent occurrence as denominatives. One such pattern has already been discussed, Fa88aL-, which expresses with various modifications the making or doing, or being occupied with, the thing indicated by the noun from which it is derived.

ruxa:m	marble	<	*raxxam-*	to pave with marble
jild	skin	<	*jallad-*	to bind a book with skin
xaymah	tent	<	*xayyam-*	to pitch a tent
qaws	bow	<	*qawwas-*	to bend
jayš	army	<	*jayyaš-*	to collect an army
šarq	east	<	*šarraq-*	to go east
ġarb	west	<	*ġarrab-*	to go west (35)

Besides, let us suppose for the sake of argument that this process was extremely restricted in the past, or even non-existent. Does it follow that we should not resort to it in the present time, especially if we know that to ban such a practice would be to create a formidable obstacle in the way of expressing an increasingly large number of concepts, particularly in the field of science and technology? It is quite inconceivable that the language could cope with the ever-growing demands if such restrictions continue to be imposed on it.

By the way, we also find it equally unjustifiable that this process should not be permissible outside scientific language. Literary Arabic, whether Old or Modern, by no means lacks instances of noun-based derivations. One of the well known proverbs of early Arabs goes: *ʔinna l-buġa:ṭa biʔarḍina: yastansiru* "the kite in our land becomes a vulture",(36) where the concrete noun *nasr* "vulture" is laid at the basis of the verbal derivation *yastansir-* "to become vulture-like".

From a practical point of view, derivation from concrete nouns is advantageous in that it enables us to avoid long and circumlocutory expressions resulting from the adoption of roundabout methods involving the use of the concrete nouns themselves. For example, a phrase such as *tasakkuru n-naša:?* "amylolysis" (where *tasakkur* < *sukkar* "sugar") is comparatively shorter while no less understandable than *taḥawwulu n-naša:?i ?ila: sukkar* "the conversion of starch into sugar". The same argument holds true in the case of derivation from borrowed (arabicized) concrete nouns, a process widely observable in the scientific vocabulary of SA. Consider the following instances:

šira:bun mukarban carbonated beverage	vs.	*šira:bun mušabbaᶜun bi (ṯa:ni: ?uxsi:di l-karbo:n — a beverage* impregnated with carbon (dioxide)
at-tamaġnuṭ magnetization	vs.	*at-ta?aṯṯuru bi'l-maġnaṭi:s* being influenced by magnet
al-hadrajah hydrogenation	vs.	*al-muᶜa:lajah bi'l-hi:dro:ji:n* treating with hydrogen

Again, a favourable factor in the application of the above method is the fact that there are in Arabic certain morphological patterns which can be of great help in this respect. The examples just cited, for instance, are cases where processes of change, conversion or conditioning are involved. A pattern which seems to be adequately capable of denoting the foregoing processes is the verbal taFa88aL-, whose corresponding nominal pattern is taFa88uL. To substantiate the preceding statement, here are some examples already current in the language: *tajallad-* "(of water) to turn into *jali:d* 'ice'"; *taṣallab-* "(of something soft) to become *ṣulb* 'hard, stiff'"; *tanaṣṣar-* "to become *naṣra:niyy* 'a Christian'"; *taḥaḍḍar-*"to become urbanized or civilized" < *ḥaḍar* "a civilized region or town".

Derivation from concrete nouns may thus be described as a direct and economical method which can be, or rather is, indispensable in scientific as well as literary language.

As far as other languages are concerned, derivation from concrete nouns has been a very useful and widely employed means of lexical growth. In English, for instance, derivatives of the type represented by "computerize" are continually being increased, specially in scientific and technical terminology. Even proper names have sometimes been used for a similar purpose. Thus the name of the French chemist Louis Pasteur (1822-1895) has occasioned the occurrence of such words as "pasteurize" and "pasteurization". The same may be said about "boycott", which traces its origin back to the nineteenth-century land agent Charles C. Boycott, who in 1880 was ostracized by the Irish Land League for refusing to reduce rents. This word has become so useful that it has passed into several European languages, e.g. French, Dutch, German.

(ii) Nouns of instrument (?Asma:?u l-?A:lah)
Another issue around which academicians engaged themselves in long and heated debate is that concerning the formation of nouns denoting instruments. Much of the controversy has revolved around the question of whether analogical derivation of such nouns should be confined to triradical transitive roots only (which constitutes the view of classical philologists) or also be applied to intransitive verbs as well as concrete nouns (as is advocated by some more recent scholars). There are also a number of other important issues which we shall need to discuss in due course. Meanwhile we find it plausible to quote the relevant resolution issued by the Cairene Academy at the end of its discussion of the subject over several sessions. The resolution in question reads (MML^{c}AM, vol.i, 1936, p.397):

> Derivatives are analogically formed from triradical verbs according to the patterns miF8aL, miF8a:L and miF8aLah for the denotation of an instrument by which something is manipulated. The Academy also decrees that the patterns of instrumental nouns [already] in common use be adhered to. [Otherwise] if no such pattern has been formed from a verb, this may be performed in accordance with any of the three patterns previously mentioned.

One thing to be remarked about the above resolution is that it has avoided any specific reference to the controversial issue mentioned on this page. This is obviously due to the sharp division of opinion characterizing the discussions that preceded the issuing of this resolution. Thus, on the one hand there are those who contend that only transitives admit of the analogical application of the aforementioned patterns; others argue that intransitives, too, are capable of such formations.

We have noticed, however, that this discrepancy is due to the fact that writers have not been unanimous in their conception of what constitutes an *?a:lah* "instrument", and this it seems has some reflection upon their views concerning whether or not intransitives should be considered as permitting analogy in the present case. Thus, from the point of view of those who define an *?a:lah* as a means whereby something is performed, e.g. *mibrad* "file", *mifta:ḥ* "key", *miṭraqah* "hammer", etc., it would seem plausible to argue that transitive roots or verbs should underlie such derivations. For an action-object relationship is a concomitant feature of any process involving the use of an instrument. The examples just cited serve to illustrate our preceding statement.

According to others, however, the concept of *?a:lah* has a far wider application; it is taken to mean anything that can serve as a means to an end or is somehow related to the achievement of a certain act. With this view in mind, al-Magribiyy, for instance, cites as nouns of instrument deriving from intransitive verbs, cases such as (ibid., p.387):

minxar	nostril	<	*naxar-*	to snort
mi^{c}ra:j	ladder	<	*^{c}araj-*	to rise, ascend
midxanah	chimney	<	*daxan-*	to smoke (v.i.)

A comparison of the two sets of examples given above is bound to reveal the fact that the concept of *ʔa:lah* as defined earlier is not equally applicable in both cases. The latter examples do not exhibit a similar action-object relationship to that exhibited by the former. It follows then that unless this concept is modified so that it covers the latter cases, these should be considered as representing a different concept, viz., nouns of place. The reason why they have not been designated as such seems to be because they happen to pattern like nouns of instrument. The writer we have just quoted, for instance, argues that although the word *mimlaḥah* indicates a container for salt, *milḥ*, it is not a noun of place since it assumes the pattern miF8aLah rather than maF8aLah (ibid., p.388), the latter being the one usually found with nouns of place or where something is found in abundance (see Appendix). Thus instead of being described as nouns of place patterning like nouns of instrument, such cases are often mistakenly treated as denoting instruments. An important factor accounting for such phenomena seems to consist in the fact that the confusion traces back to early grammarians whose works have always been taken for granted and are rarely called in question by later scholars (e.g. Wāli, 1936, pp.371-8). In his account of the subject, Sībawayh (1361 AH, vol.2, p.249) for instance treats *miḥlab* "milk-pail" on a par with *minjal* "scythe", *miqra:ḍ* "scissors", *miksaḥah* "shovel", etc., all as instances of "what you manipulate with" (*ma: ᶜa:lajta bih*), i.e. instruments. Here are some more examples of early nouns of place that have been mistaken for nouns of instrument:

miᶜlaf	feeding-place, fodder-bag
miʔba:r	needle-case
miḥbarah	ink-pot
miṣdaġah	pillow
mirfaqah	cushion to lean upon

According to classical philologists, nouns of instrument should be exclusively derived from the first form of the verb, which must be transitive in signification.(37) It follows then that other (derived) forms of the verb are not considered as permitting analogy, although they may be transitive. It is on this basis that a neologism like *midfaʔah* "stove" would be objected to; for the corresponding transitive verb in this case is either the second (derived) form Fa88aL-, i.e. *daffaʔ-* or the fourth ʔaF8aL-, i.e. *ʔadfaʔ-* "to warm, heat", while the ground forms *dafiʔ-* and *dafuʔ-* "to become or feel warm" are both of intransitive signification. (There happens to exist no verb of the form **dafaʔ-*.)

In actual linguistic practice, however, the neologism quoted above, as well as others representing the same case, do exist in the language. It seems that in such cases, where the idea expressed by the root is capable of transitive connotation, it is irrelevant to base our analysis on what particular form the corresponding verb has, since the nature of the pattern involved allows the representation of only the root radicals, i.e. in addition to the pattern-constituting elements. A word like *miḥra:k* "poker", which derives from ḥ-r-k and where the corresponding transitive verb is *ḥarrak-* and not **ḥarak-*, would thus be analysed into the root constituents ḥ,r,k and the pattern miCCa:C (where the Cs represent the respective positions of the radicals). A different pattern, on the other hand, may allow of the representation of other additional elements. muFa88iLah, for instance, gives us *mubarridah* "cooler" where the second radical is doubled as in the corresponding second (derived) form *barrad-* "to make cold" from *barad-* or *barud-* "to be or become cold" b-r-d-.

Another point to be observed about the resolution quoted above concerns the concentration, as analogically applicable, on the patterns miF8aL, miF8a:L and miF8aLah to the neglect of a number of others that have been operative in the language or play an increasingly significant role in the configuration of present-day scientific vocabulary. It is obvious that the academy's intention has been to keep the language as homogeneous as possible. Early investigators of the langauge have observed the high frequency of these patterns with nouns of instrument, and modern academicians in their turn must have felt it their duty to stick to, or at least favour, these patterns so that the homogeneity just alluded to may not be disturbed. However there are a number of relevant facts which seem to have been overlooked and which therefore call for some consideration.

There is no question about the plausibility of harmony and homogeneity in language; nothing would be more convenient for the speaker than to have a limited set of commonly specifiable morphological patterns in accordance with which he may be able, say, to increase his repertoire of instrument-denoting terms. But how far could this be carried out, or rather how far would language lend itself to be so fixedly controlled? And would there be any justification for hindering the language from following what may be a natural course of development, especially if this represents an attempt on its part to cope with new demands.

To start with, the criterion of majority (*kaṯrah*) vs. minority (*qillah*) on which so many philologists have based their argument for the analogical vs. non-analogical categorization of the native vocabulary has unfortunately almost always been employed for prescriptive rather than descriptive purposes, which at times has resulted in its being more of a hindrance than a help. Its influence has been such that it has restricted the freedom of present-day scholars (specifically translators and other people engaged in the process of creating new vocabulary) to adopt patterns that have been as commonly and as happily adopted as any others. Since these are all products of the native tongue and have proved to be of equal service to the speakers, it is hardly conceivable that their adoption should be checked or restricted. Besides, the frequency or infrequency of linguistic phenomena in early Arabic cannot actually be said to have been accurately and precisely surveyed by the early codifiers of the language — at least not always. It is reported, incidentally, that nouns denoting the concept of instrument in ancient Arabic were more frequently of the pattern Fi8a:L than of any of the three others mentioned above.(38)

It has already been implied that while paying much attention to the views and rulings of early philologists, contemporary studies on Arabic in general have failed to investigate the developmental aspects of the language and its trends of natural growth in the present circumstances. To take the topic at issue as a case in point, we thus find that while discussions are almost exclusively focused on the triad of miF8aL, miF8a:L, and miF8aLah, very little attention is paid, if any, to the fact that present-day Arabic shows a particular tendency towards using active participles for the denotation of instruments and other objects whose function is underlaid by the same concept – this being the case although many such applications are highly frequent even in the dictionaries and glossaries issued by the academies themselves (as we shall see later). Besides, the development in question does not constitute a totally novel phenomenon in the history of Arabic. Al-Rāziyy (d.932 AD) (1964, p.8) tells us for instance that of the instruments commonly known by his contemporary scientists there were *al-ma:sik* Fa:8iL, "an instrument for holding or gripping things", and *al-muqattiᶜ* muFa88iL "cutter, shears". Other early instances include: *za:jil* "wooden stopper of a skin bag", *ja:miᶜah* "chain", *ra:mij* "decoy-bird", etc. (see al-Athariyy, 1963, pp.21-2).

That the occurrence of active participles as nouns of instrument is highly frequent in present-day Arabic is indicative of the evolutionary aspect of their denotative potential. One thing which both active participles and instruments have in common is that they both play the role of active participant in the process involving their employment. Besides, Arabic active participles, especially when derived from transitives, are "not only real participles, indicating a temporary, transitory or accidental action or state of being, but also serve as adjectives or *substantives, expressing a continuous action, a habitual state of being or a permanent quality*" (emphasis mine) (Wright, 1967, vol.1, pp.131-2). Instruments may similarly be described as being designed or set to be permanently associated with certain functions or purposes. A "key", for instance, has the permanent characteristic of being an instrument of opening; a "lancet" of cutting; a "drill" of boring, and so on. The following are instances of the occurrence of active participles as nouns of instrument in the scientific vocabulary of SA. Many of these instances have already been entered in scientific and technical dictionaries issued by the Egyptian (see Arabic Language Academy, 1957), Iraqi (see Iraqi Academy, 1977), and Syrian (see Syrian Academy, 1977) academies. The Iraqi and Syrian publications are both issued under the auspices of the Arab Organization of Education, Culture and Sciences.

Fa:8iL, Fa:8iLah <	Form I	
la:qiṭ	pick-up	radio, tv
ġa:liq	shutter	photography
ša:ḥin	charger	elec. engineering
ka:biḥ	brake	engineering
ka:bisah	press	engineering
ᶜa:kisah	reflector	physics
ra:?iyah	viewfinder	photography
?a:latun ka:tibah	typewriter	typography

muFa88iL, muFa88iLah < Form II

muraššiḥ	filter, percolator	engineering
mujaffif	desiccator	physics, chemistry
muqawwim	rectifier	elec. engineering
mujassim	stereoscope	photography
mukabbir	magnifier	radio, photography
mukat̲t̲ifah	capacitator	elec. engineering
mubarridah	cooler	mech. engineering
mujammidah	freezer	mech. engineering

muFa:8il, muFa:8iLah < Form III

muqa:wim	resistor	elec. engineering
muᶜa:dil	equalizer	elec. engineering
mufa:ᶜil	reactor	elec. engineering, physics
muḍa:ᶜif	multiplier	elec. engin., chemistry
muza:min	synchronizer	radio
muqa:rin	comparator	physics
muᶜa:dil	compensator	elec. engin., photography
muha:yiʔah	adapter	engineering

muF8iL, muF8iLah < Form IV

muskit	silencer(39)	mechanics
mu:hin	attenuator	elec. engineering
mušᶜir	indicator	engineering
mursid masa:r	track guide	radio, aeronautics
mursilah	transmitter	elec. communications
al-mubriqatu l-ka:tibah	teletype	telegraphy

mustaF8iL, mustaF8iLah < Form X

mustaqṭib	polarizer	physics
mustaxliṣ	extractor	engineering
mustaqbilah	receiver	elec. engineering
mustaḥlibah	emulsifier	chemistry
mustaxliṣ	demodulator	radio

In addition to the above patterns, there are also Fa88a:L and its feminine counterpart Fa88a:Lah. These two patterns are traditionally known as intensive adjectives (or nouns), corresponding to the verbal adjectives Fa:8iL and Fa:8iLah. For example *kaḏḏa:b* and *kaḏḏa:bah* "a habitual liar" are the intensive forms of *ka:ḏib* and *ka:ḏibah* "lying". An easily detectable development of modern SA is the highly frequent occurrence of these patterns as denoting instruments, devices and machines that perform something, or by means of which something is done, regularly and constantly.

The following are only a few of the instances we have been able to attest:

ᶜadda:d	counter	data processing
hazza:z	vibrator	physics
raqqa:ṣ	pendulum	physics
saffa:ṭ	aspirator	physics
qalla:b	tip-cart	civil engineering
xalla:ṭ	mixer	metallurgy
xarra:mah	punch	engineering

ḥaffa:rah/baḏḏa:rah	drill	mining/agriculture
haṣṣa:dah	mowing machine	agriculture
samma:ᶜah	stethoscope	medicine
ᶜaṣṣa:rah	squeezer	
kassa:rah	nutcracker	
farra:mah	mincing machine	
ġassa:lah	washing machine	

It is hoped that the point has been made clear that attempts aimed at assessing or promoting the viability of the language at a certain stage should not be based solely on the set of facts or observations established at previous stages of its history, let alone the fact that these need not always be taken for granted. Besides, our loyalty to our language need not imply that we should oppose or ignore its developmental tendencies, especially if these represent an attempt on its part to follow its natural course of growth. (For further features of lexical development and adaptation, see section 2.4.2.)

It should be understood, however, that our preceding statements are not intended to imply that already established patterns, e.g. miF8aL, miF8a:L and miF8aLah have ceased to be operative or are playing a less significant role in the configuration of present-day Arabic vocabulary. Our intention has been to point to the current tendency of the language to utilize other available resources which represent its potential capacity for development. It may well be useful to cite some examples of recent coinages modelled on the patterns just mentioned:

misᶜar	calorimeter	physics
mišbak	clasp	design engineering
mirqab	telescope	optics
miṣᶜad	lift	mechanical engineering
mijhar	microscope	optics
miḏya:ᶜ	radio-set	radio
miġwa:z	gasometer	chemistry
mixma:r	vinometer	chemistry
minġa:m	tonometer	acoustics
minfa:s	spirometer	medicine
mirwaḥah	propeller	engineering
mirjafah	tremograph	geophysics
mirsamatu ḏ-ḏabḏaba:t	oscillograph	elec. engineering

2.4.2 Views and proposals

In the preceding section we have tried to provide the reader with an idea of how the question of lexical creation through analogical derivation (as exemplified by certain cases) has been approached in academic circles. One of our main objectives has been to reveal the sort of impact these approaches are likely to have, or have already had on the language. Now it is time we turned to the situation outside the context of academies, though throughout the course of our investigation some reference may occasionally be made to certain academic viewpoints.

(i) The effort of Arab writers and scientists to enrich their language with new adequate scientific vocabulary has manifested itself, among other things, in their attempt to specify certain patterns which hitherto have not been so specifically distinguished as designative of particular meanings. Due to the fact that more than one pattern may be potentially capable of expressing more or less the same idea or meaning component, different writers have employed different patterns for the same signification. Let us take as a case in point a number of attempts aimed at furnishing Arabic equivalents for foreign terms of the type ending in the adjectival suffix -able or -ible – that is, words expressing the idea that what is denoted by them is capable of, fit for, characterized by or worthy of being so acted upon or towards; e.g. permeable, combustible, notable, etc.

As far as Arabic is concerned, the meaning of the type of words just exemplified is usually expressed by means of set combinations or more-than-one-word lexical units consisting of the adjective *qa:bil* "capable, etc.", followed by the preposition *li-* "of, to, for", plus a verbal noun governed by the latter. Sometimes the constituent *qa:bil* is replaced by one or more of a variety of equivalent words like *ṣa:liḥ* "fit for, appropriate", *ᶜurḍah* "liable". etc. Other types of syntactic expressions are also used for the same signification. Consider the following instances:

qa:bilun lin-nufu:ḏ	permeable
qa:bilun lil-ʔiḥtira:q	combustible
ṣa:liḥun lil-ʔakl	eatable
ᶜurḍatun lit-talaf	perishable
*sari:ᶜu l-ġaḍab**	irritable
*mutaᶜaḏḏirun ʔinqa:ṣuh***	irreducible

* Literally: quickly (or easily) made angry.
**Literally: difficult to reduce.

However, these expressions are rather long (at least as far as scientific language is concerned) and do not lend themselves easily to further derivation, which accounts for the emergence of a number of alternative methods based on the principle of derivation where the Arabic equivalent is reduced to a single-word unit. For instance, in his glossary of scientific terms, al-Kawākibiyy (1947, pp.15,18,26,31,47,51) employs the pattern Fa8u:L for the expression of the signification in question. In his terminology, "permeable" would thus be rendered *nafu:ḏ*; similarly, "digestible" would be *haḍu:m*, "inflammable" *lahu:b*, etc., while "inflammability" would be *al-lahu:biyyah* as opposed to the alternative long syntactic expression *al-qa:biliyyatu lil-ʔiltiha:b* "the capability of being inflamed".

Resort to the pattern Fa8u:L seems to be accountable for in terms of the fact that it is already operative in the language with a function approximate to that of English adjectives with the -able ending. Let us take a few cases in point. When a person is described as *ṣadu:q* this implies that he is characterized by being always truthful and honest; similarly a *kaḏu:b* is one who is addicted to lying and therefore characteristically prone to it.

Fa8u:L thus expresses a state or quality permanent in the thing or person referred to. The same may be said to be more or less applicable in the case of, say, English "inflammable" (or rather "flammable" these days) which indicates that what is referred to is something which will burn readily or quickly, i.e. is characterized by this quality.

Another characteristic of this pattern is that when derived from a transitive root it may also have a passive meaning. For example, a horse which is *raku:b* < r-k-b "to ride", is one which will allow (and is thus capable of) being ridden upon; and a *ḥalu:b* "cow" (< ḥ-l-b "to milk") is one that may be milked, i.e. ready or fit for yielding milk. Here we have cases parallel to such English words as "edible" (i.e. fit to be eaten), "soluble" (i.e. that can be dissolved), "understandable" (i.e. that can be understood), etc.

An additional factor in favour of employing the pattern Fa8u:L for the above signification is that it constitutes one of no less than twelve patterns employed by Arabic to express more or less the same idea: intensiveness and inherent characteristics. Since it is not the most commonly used of these patterns,(40) it seems plausible that it be assigned the specialized meaning at issue, at least as far as scientific language is concerned.

It might be argued that the use of this pattern in this manner might give rise to ambiguity in some cases. For instance, there already exists a word *ʔaku:l* < ʔ-k-l "to eat", which is an intensive form meaning "(one who is) voracious" – and so, the argument may go, this meaning is likely to be confused with the new one "edible, eatable". Such an argument, however, is not hard to refute. First, the context in which such a word occurs is bound to render it unambiguous; secondly, in a phrase like *ma:ddatun ʔaku:lah* "edible substance", it would be false to claim the existence of any confusion or ambiguity since the noun accompanying the adjective is inanimate, whereas eating is a characteristic peculiar to animate creatures only. In other words, the noun *ma:ddah* here plays the role of a patient rather than that of an agent. Conversely, the noun in *fata:tun ʔaku:lah* "voracious girl" is clearly agentive.

In view of the facts stated above, we incline to the opinion that the pattern Fa8u:L is a useful and economical method of expressing the meaning under discussion, especially for scientific and technical purposes. Following this method we could thus say:

kasu:r	breakable
ḥalu:l	analysable
ḥaru:q	combustible
ṣahu:r	fusible
kaṯu:f	condensable
talu:f	perishable
xaṯu:r	coagulable
xaṯu:riyyah	coagulability
ḍaġu:ṭiyyah	compressibility
šaᶜu:liyyah	ignitability

etc.

The pattern Fa8u:L, however, has not been the only case in point. Muṣṭafa Jawād (1955, p.108), a member of the Iraqi Academy, argues that the active participle muF8iL, derived from Form IV, i.e. ?aF8aL-, has been used by the Arabs in a way similar to that involving the use of English and French words ending in -able or -ible. Thus, to quote his examples, there are *muṯmir* "bearing fruit", *mu:riq* "producing leaves", *munjib* "begetting or bearing children", etc. In an article published in the journal of the Syrian Academy (1957, pp.133-4) the same writer proposes another form mustaF8iL < Form X, i.e. istaF8aL- for the same purpose, e.g. *mustaḥṣid* "harvestable". Some such coinages have already been adopted by the Iraqi Academy. There are for instance *mustanqil* "transmissible", *mustaḥrik* "movable", *mustamṭil* "ductile" etc. (Iraqi Academy, 1967, pp.10,7,4 respectively).

While not denying the possibility of using such a pattern, we find it more economical to employ the previously mentioned Fa8u:L, which is obviously much simpler and shorter. The above examples may thus be rendered *naqu:l, ḥaru:k, maṭu:l*.

Another similarly used pattern is Fa8i:L whereby the Egyptian Academy has rendered the foregoing example "ductile" into *maṭi:l*, thus adopting a different pattern from that adopted by the Iraqi Academy for the same word (see The Permanent Bureau of Arabization, 1971, p.45). The Egyptian Academy has also resorted to another method of a more or less different nature. This consists in the prefixing of the definite article to the passive form of the imperfect verb, e.g. *al-yu?kal* "edible". This method, however, has hardly gained any acceptance.

In his dictionary of scientific and technical terms, al-Khaṭīb (1978, p.111) gives the form *munḍaġiṭ* munFa8iL as an equivalent of "compressible". It should be remarked, however, that this is an inadequate rendition, for the passive participle munFa8iL indicates a perfective aspect and expresses a non-habitual action, an action considered from the standpoint of its completion or conclusion. In other words, *munḍaġiṭ* is a counterpart of "compressed" rather than "compressible"; whereas adjectives with the suffix -ible/-able express such qualities as susceptibility, preparedness, capability, etc., as being characteristic of and always inherent in the thing or person they refer to. Again, the pattern Fa8u:L seems quite eligible for the expression of such notions, and so the word "compressible" would more adequately be rendered *ḍaġu:ṭ*.

We have thus seen that by way of attempting to furnish appropriate scientific vocabulary Arab scholars have resorted to no less than five lexical patterns (Fa8u:L, muF8iL, mustaF8iL, Fa8i:L, munFa8iL, etc.) whereby to achieve more or less the same end. This state of affairs serves as clear indication of the adaptational movement taking place in present-day Arabic vocabulary. Students of science who use Arabic as their medium, or at least aspire to, clearly have realized the urgent need for their terminology to be as economical and practical as possible. From another point of view, the same state of affairs may also be described as indicative of the great potential of Arabic to evolve and grow from its own essence.

(ii) What has been said above concerning the assignment of morphological patterns to certain specialized meanings could be said to represent one aspect of a large-scale process of adaptation and growth now being observed in scientific Arabic. The instances of patterns we have just discussed are all already occurrent in the language, but nowadays are being adapted to certain specific significations in the language of science. Another aspect of this process consists in the fact that there have been several attempts on the part of modern Arab scholars to investigate their lexical heritage with the intention of reviving further patterns that were operative in earlier Arabic and which may now be seen as being potentially capable of enriching the language with new scientific vocabulary. These scholars must have realized that the number of patterns already operative in the language is not sufficient enough to meet the large variety of concepts brought about by scientific and technological advance.

Thus al-Shihābiyy finds that the pattern muFa8LaLah was used by early Arabs for places which abound in different kinds and species of plants, animals, etc., whose names consist of more than three radicals. Early instances of words assuming this pattern are cited in Ibn Sīdah (n.d., vol.4, sec.14, p.205) (d.1066), e.g. *?arḍun mu?arnabah* is a land abounding in rabbits, *?ara:nib* (pl. of *?arnab*). Al-Shihābiyy argues that the pattern in question could be employed as a counterpart of such French suffixes as -aie and -erie which serve a similar purpose. In this way, he goes on, we could coin words like (1965, vol.xl, pp.366-8):

muṣanbarah < *ṣanawbar*		pine trees	Fr. pineraie
muzaytanah < *zaytu:n*		olive trees	Fr. olivaie
muṣafṣafah < *ṣafṣa:f*		willow trees	Fr. saulaie

A similar attempt is made by Ghazāl (1977, p.53) whose investigation has reavealed the patterns:

Fu8LuL & Fu8Lu:L	for names of birds, insects and small mammals
Fi8La:L	for large animals
?iF8i:L	for names of instrument
Fa:8u:L	for names of plants, rocks and tools

Another scholar, al-Sāmarrā?iyy, reminds us of a further ancient pattern, Fi8a:L (see fn.38), which he finds more convenient for the denotation of instruments than currently used ones including those that are analogically derived. He also advocates the utilization of Fu8Lah which represents the ancient form of passive participle used in Arabic prior to its development into the now current maF8u:L, e.g. *ḍuḥkah* "laughing-stock", as opposed to *maḍḥu:kun minh* "(one who is) laughed at" (see al-Sāmarrā?iyy, 1977, p.157).

As far as actual linguistic practice is concerned, proposals of the type just exemplified have so far met with little if any success. One factor is that such attempts fail to take into account the fact that what was applicable in the past need not always be so in the present. That a certain pattern was operative or productive at an earlier stage of the language need not necessarily constitute sufficient incentive for the current speakers to re-adopt it, let alone the fact that they may not even be aware of it, which is indeed mostly the case with such patterns. It is an established fact that a decisive factor in the currency of any linguistic form or, in this case, pattern, is its conventional acceptance by those among whom it is in current use. Words and the patterns they assume are not much different in this respect from coins. No matter how valuable the latter are in their respective eras, they are bound to lose that value once they have been displaced by differently minted ones. It is no coincidence then that terms such as "coin" and "mint" have come to designate both types of process.

It is also relevant to point out here that, contrary to a widely held view, morphological patterns are not immune against processes of change or development (see fn.38). We may add here that such changes were noticed and recorded by very early philologists. Over a thousand years ago, al-Kisā?iyy, for instance, observed that his contemporaries showed a tendency towards modifying the pattern Fu8Lu:L, as in *ṣuᶜlu:k* "beggar", into Fa8Lu:L. The crux of the matter, however, is that such phenomena, no matter how extensive, are most often not regarded as developmental features but rather as instances of incorrect speech, as al-Kisā?iyy himself does (see Ḥijāziyy, 1970, pp.60f.).

(iii) Attempts at meeting the growing need for more patterns have not been confined to the revival of old ones; there has also been a tendency on the part of a few scholars to adopt what may be described as innovatory approaches to the problem. The main proposals put forward by such scholars are:

(a) utilizing certain morphological features of Arabic and developing them into more productive means of lexical creation;
(b) extending the denotative applicability of already existing patterns and generalizing the analogicality of certain morphological features: and
(c) inventing entirely new patterns and associating them with certain designatory functions.

The following discussion investigates the practicability and validity of such proposals.

Type (a) proposals may be represented by Tammām Ḥassān's attempt which is based on the concept underlying the employment of the so-called *ḥuru:fu z-ziya:dah* "letters of addition". It has been mentioned above that Arab grammarians have identified ten of the Arabic letters as being capable of effecting certain modifications on the meaning of a form when used in addition to its radical constituents (see footnote 9 to Chapter 1). Ḥassān's proposal consists in that not only these, but all the letters constituting the Arabic alphabet, are capable of playing the same role. He thus argues that the letter /d/, for example, if added to, say, the already existing Fa8aL-, would give us the new patterns daF8aL-, Fad8aL-, Fa8daL- and Fa8Lad; each of which may be assigned a certain signification. Applied actually to the verb *saxan-* "to become hot", such a process would result in the creation of a new set of verbs consisting of *dasxan-*, *sadxan-*, *saxdan-*, and *saxnad-*, which (the writer goes on) may be used to yield further derivatives such as adjectives, participles, verbal nouns, etc. (see Ḥassān, 1973, pp.153-4).

Theoretically speaking, this proposal seems very attractive; in practice, however, it is bound to involve a high degree of impracticability. The fact that the patterns yielded by this method have never occurred in the language before would inevitably present the speakers with the heavy burden of associating the new patterns with the new meanings or concepts conveyed by the lexical items assuming them. Such a view can thus be objected to on the grounds of regarding language as an end in itself rather than a means to an end.

For a representative example of proposals belonging to group (b) above, we may take al-ᶜAlāyli's attempt as a good case in point (1940, pp.53-96). Seeing that Arabic is basically a derivational language, and noticing at the same time the serious challenges facing the language in its present time, he advocates as a means of reformation, the application of the existing patterns of Arabic to a wider field of designation. Such a procedure is seen as a protective measure against the adoption of word-formational processes that are not characteristic of the language. Thus he suggests for instance that the pattern Fa8La:? which is traditionally found with words denoting feminine nouns and adjectives, e.g.*ḥasna:?* "a beautiful woman", *ṣafra:?* "yellow", etc., be used for the designation of places with certain features that are characteristic of them. For example, a place where there is much industry, according to al-ᶜAlāyli, would have the name *ṣanᶜa:?* from ṣ-n-ᶜ "to manufacture". Similarly Fu8La:n which is normally assumed by verbal nouns, e.g. *šukra:n* "thankfulness", and adjectives, e.g. *ᶜurya:n* "naked", would be used to signify the source or starting-point of the thing referred to, e.g. *nuhra:n* "the point at which a river (*nahr*) starts".

It is obvious that the link is completely lost between the original signification of the patterns cited above and the new meanings they are intended to stand for. In other words, cases such as the ones in question cannot be said to represent a natural process of development whereby the denotative range of the patterns concerned is extended to cover wider areas of signification. Rather it is a prescriptive method requiring the conscious application of these patterns to meanings with which they have not been associated before. It should not be surprising then that proposals of the type under discussion have met with no success whatsoever.

Another proposal to be included in the present group is that of a Moroccan scholar, Ghazāl, who in a discussion with the author advocated the employment of the termination *-ya:?* as a counterpart of the combining form "-logy" found in such English words as "geology", "psychology" etc.(41) This Arabic termination is already exhibited by a small number of words denoting the names of sciences, e.g. *ki:mya:?* "chemistry" and *fi:zya:?* "physics", which as we have seen can be traced back to the early stages of the first scientific renaissance. It should be mentioned, however that the element *-ya:?*

has not so far been productive; there have been produced no further forms terminated by this element and where the function is to denote a scientific field or discipline. Nevertheless, the adoption of this proposal seems to be justifiable on a number of grounds. First, the occurrence of this termination in new forms is not going to be received by the speakers as an entirely novel phenomenon since instances already exist in the current vocabulary. Secondly, this method is useful in that it helps avoid the use of long periphrastic expressions such as *ᶜilmu ṭabaqa:ti l-?arḍ* "the science of the strata of the earth", i.e. geology; whereas *?arḍiya:?* from *?arḍ* "earth" constitutes a far more economical term. Notice also that the latter, unlike the former, presents us with no difficulty when it comes to deriving further forms, e.g. *?arḍiya:?iyy* "geological", *?arḍiya:?iyyan* "geologically", etc. There is also the fact that, being easily distinguishable both formally and semantically, this morphological device is not likely to be a source of confusion or ambivalence.

As to proposals of type (c) above, we may take as our case in point another attempt made by al-ᶜAlāyli, who has invented a number of entirely new patterns which he has associated with certain significatory functions (1940, pp.65-7,81). Here is a sample of the patterns he has suggested:

Fa8Lat	for the indication of hypersensitivity or hyper-irritability
Fa8Lu:t	for indicating transformation from one state to another
Fa8aLa:?	for indicating capability or capacity
Fa8aLayya:	for indicating penetration or permeation
taF8aLu:t	for indicating the instantaneous occurrence of something

Despite the fact that the patterns mentioned above were put forward more than forty years ago, they have failed to gain any acceptance or currency in actual linguistic practice. Factors accounting for this have already been implied in our discussion of the previous cases. The general remarks which follow are also relevant.

By way of concluding our investigation of the issues raised in section 2.4, it is useful to bring into focus a number of the most important facts relevant to the question of reviving or enriching Arabic vocabulary through the native method of analogical derivation.

There is no denying that lexical creation through the method of *?ištiqa:q* represents the natural and most productive means Arabic has adopted over its history of development. It is also undeniable that the practice of forming new vocabulary in accordance with native morphological patterns is equally useful in that it preserves the homogeneity and characteristic structure of the language. However, attention must be drawn to the fact that the usefulness of these means can be maintained only so far as they are allowed to follow their spontaneous nad natural trends of growth, in which case adaptational and developmental phenomena will not lack the capability of practicable application. An important fact to be taken into account is that what is theoretically feasible need not necessarily be always so in actual practice. It follows then that prescriptivism is not a workable procedure in language. Evidence for this statement has already been drawn from the fact that proposals for reform which assume this nature have failed to achieve any success. Mention must be made of the fact that some scholars seem to form their views against a background far removed from what is actually taking place in real life, which corresponds to another fact – namely, that their efforts are often characterized by their tendency to think of language as an end in itself rather than a means to an end.

What has been said above, however, need not imply that all efforts aimed at enhancing the process of lexical expansion are doomed to failure. We have seen that success has not been lacking in cases where such efforts are based on the utilization of the potential tendencies of the language and the natural trends of linguistic evolution. In such cases, one would feel inclined to go even so far as to argue that prescriptivism or some form of it is not totally objectionable since it may be seen as a sort of measure whereby to regulate or promote a certain phenomenon which is already in the process of crystallization.

Notes to Chapter 2

1 For an account of this type of derivation and the definition just quoted, see the following writers: al-Suyūṭiyy (n.d., vol.1, p.346); Ibn Durayd (1958); al-Sayyid (1976, p.42); al-Shiḥābiyy (1955, p.10); al-Mubārak (1960, pp.55ff); Fahmi (1961, pp.328f); Ḥassān (1974, p.178); Wāfi (1972, pp.178f); and Tarzi (1967, pp.1-19).

2 These two types are also known as *al-ʔištiqa:qu l-kabi:r* "large derivation" and *al-ʔištiqa:qu l-ʔakbar* "major derivation", respectively.

3 See Ibn Jinniyy (1913, vol.1, pp.525-31).

4 It is obvious that this claim, which in modern linguistics would be treated under Sound Symbolism, is open to criticism. However, it is not our concern here to give a critical account of this theory; for such an account the reader may refer to Tarzi (1967, pp.323-35).

5 See Wāfi (1972, pp.184-6).

6 For additional information on both *al-qalb* and *al-ʔibda:l*, the reader may refer to al-Mubārak (1960, pp.67-91).

7 In its widest sense *qiya:s* means "generation" and may thus be applied to other domains of language, e.g. syntax, where the implication is to generate new grammatical sentences depending on the existing rules of grammar. It is also applicable in the case of loanforms that have been fully assimilated to the native morphological system and thus made analogous to indigenous formations; hence our term "Analogical Arabicization" (Chapter 4).

8 According to M.A. Mazhar (1963, p.50), Arabic is "so systematic, scientific and philosophic that it must have come from the hand of God".

9 For further examples of this sort and a discussion of this phenomenon, see Sheard (1954).

10 This, however, need not imply that pre-Islamic Arabs were totally unfamiliar with any aspect of human civilization, nor that their language was lacking in the inherent capability of abstract expression. See al-Sāmarrā?iyy (1977, pp.27-38) and ᶜAbd al-Raḥmān (1971, pp.41-51).

11 For more examples of early Islamic terminology, see al-Suyūṭiyy (n.d., vol.1, pp.294-303); Ibn Fāris (1963, pp.79-81; and Zaydān, "al-Lughah..." (n.d., pp.64-6). The latter also cites a number of words associated with pre-Islamic concepts but which, due to the sociocultural changes brought about by Islam, soon began to drop out of use and sink into oblivion (pp.66-8).

12 For an account of these schools, their views and initiators, see: al-Makhzūmiyy (1955); al-Makhzūmiyy (1974); Tarzi (1967, pp.57-66); al-Suyūṭiyy (1326 AH); Ibn al-Anbāriyy (1364 AH),etc.

13 These were one of the oldest sects of Islam. They rebelled against ᶜAli, the Prophet's cousin and son-in-law when, on the field of Ṣiffīn, he consented to Muᶜāwiya's proposal that their claims to the Caliphate be decided by arbitration. They are well known for their piety, fanatical adherence to their theories and the utmost severity with which they practised their principles. Their most important teachings were that a guilty caliph must be deposed or even put to death, and that every free Arab has the right to be a caliph.

14 See, for instance, al-Khuwārizmiyy (d.997) (1895, p.5); also, al-Tahānuwiyy (1963, p.79 fn.1).

15 For a more detailed account of these and other sciences, the terminology related to them and the role of science in Islamic civilization in general, see: Nasr (1968); al-Farābiyy (d.950) (1931); ᶜAbd al-Raḥmān (1977); Zaydān, "Ta?rīkh Adab al-Lughah..." (n.d., esp. vol.2); and al-Khuwārizmiyy (1895).

16 See al-Khuwārizmiyy (1895, pp.364, 264-5).

17 See al-Rāziyy (1927, pp.412-17, and pp.345-61 for an English translation).

18 See al-Rāziyy (1964, p.11).

19 See al-Rāziyy (1964, p.11): "*wa l-qana:ni: yuḥta:ju laha: fi: taxni:qi l-muṣaᶜᶜada:ti wa ġayrih. wa ḏa:lika ?anna ?ahla l-ḥikmati rubbama: ?ara:du: taxni:qa šay?in minha: fa-jaᶜalu:hu fi:ha: wa ṣa:ᶜadu:hu wa yataxannaqu fi: ṣadriha:*".

20 Works similar to that of al-Khuwārizmiyy were subsequently carried out by many scholars: Ibn ᶜArabiyy (d.638 AH)(1972); al-Jurjāniyy (1357 AH); al-Ḥusayniyy (d.1094 AH) (1974); al-Tahānuwiyy (d. 12th century AH) (1963); Sprenger (1854).

21 According to the doctrine of the Shiites, the Imam is one of the descendants of Caliph ᶜAli, who is the divinely appointed, sinless and infallible deliverer of the Muslims.

22 See Goichon (1969, p.66).

23 al-Khuwārizmiyy (1895, p.35).

24 Schools of thought, doctrines, principles, etc. are most frequently named after their founders in the same way; e.g. *al-Yazi:diyyah (Yazi:d)*, *al-Ḥanbaliyyah (Ḥanbal)*, *al-Muxta:riyyah (al-Muxta:r)*, and in each case the term designates what constitutes a *Yazi:diyy*, *Ḥanbaliyy* etc.

25 al-Rāziyy (1343 AH, pp.494ff).

26 al-Jurjāniyy (1357 AH, pp.229, 171, 165, respectively).

27 Goichon (1969, pp.65, 62, respectively).

28 Here is an interesting occurrence of the previously mentioned suffix *-iyyah*, where the preceding element is a syntactic construction, viz. *la:?adri:* "I do not know".

29 This term is commonly used to refer to the period of Arabian history from the earliest times down to the establishment of Islam. Writers often translate it as the "State of Ignorance" or "Days of Ignorance" (see for instance Chejne (1969, p.54). It has been shown, however, that the stem *jahl* (from which *Ja:hiliyyah* is derived) as used by pre-Islamic poets is connotative of "wilderness", "savagery", rather than "ignorance". Its true antithesis accordingly is not *ᶜilm* "knowledge" but rather *ḥilm*, which denotes such moral qualities as gentleness, consideration, mildness, etc., which characterize a civilized man. See Nicholson (1953, p.30).

30 al-Thaᶜālibiyy (1938, pp.249, 250, 263, respectively).

31 al-Rāziyy (1964, pp.10, 86 respectively).

32 al-Khuwārizmiyy (1895, pp.256-7). Additional examples are provided in Appendix.

33 The history of this academy dates back to 13 December 1932 when it was first established by decree of Fu?ād I, then King of Egypt, under the name "The Royal Academy of the Arabic Language" (*Majmaᶜ al-Lughah al-ᶜArabiyyah al-Malakiyy*). In 1938 it assumed the name "Fu?ād I Academy for the Arabic Language" (*Majmaᶜ Fu?ad al-?Awwal lil-Lughah al-ᶜArabiyyah*). Then after the 1952 revolution and the subsequent abolition of the monarchy, it came to be known as "The Arabic Language Academy" (*Majmaᶜ al-Lughah al-ᶜArabiyyah*).

34 For the sake of brevity and for the benefit of translators and those concerned with the development of Arabic vocabulary, we have listed in an Appendix a number of the measures and resolutions passed and adopted by the Cairene Academy and which concern the enrichment of Arabic vocabulary by means of native-based methods, specifically analogical derivation. Examples are also provided.

35 See Wright (1967, vol.1, p.32).

36 Proverbially: he who lives amongst us becomes as honourable as ourselves. An English proverb with approximately the same idea is "our geese are all swans".

37 The first or ground form of the verb is generally transitive or intransitive in signification often according to the vowel following the first radical. Fa8aL- is mostly transitive, e.g. *katab-* "to write", but intransitive instances of this pattern are not infrequent, e.g. *jalas-* "to sit"; Fa^{c}iL- has generally an intransitive signification, while Fa^{c}uL- is invariably so. An aspect of signification distinguishing the two is that the former indicates a temporary state or condition or a merely accidental quality in persons or things, whereas the latter indicates a permanent state or a naturally inherent quality.

38 See al-Athariyy (1963, vol.x, p.20). See also the editor's comment in Zaydān "al-Falsafah..." (n.d., p.104). According to the latter, Fi8a:L is the oldest instrument-denoting pattern in Semitic. It seems that this pattern has, at a later stage of its history, undergone a process of affixation whereby the prefix *mi-* has been added to it, a process which has also incurred certain changes within the pattern itself. Evidence in support of this claim may be derived from the fact that there exist in Arabic doublets that are accountable for it. There are for example *sira:d* and *misrad* from s-r-d, both meaning "a saddler's awl or punch"; and *zila:j* and *mizlaj* from z-l-j, both meaning "a sliding bolt on a door". There also exist triplets like: *niṭa:q*, *minṭaq* and *minṭaqah*, all meaning "girdle"; and *riba:ṭ*, *mirbaṭ* and *mirbaṭah*, all with the meaning "tie, bridle". However, cf. Jawād (1955, p.20).

39 Another perhaps more current rendition of this term is *ka:tim* from k-t-m "to hide, to suppress".

40 Fa88a:L for instance is much more frequent, especially nowadays. Other patterns include: Fi88i:L, Fa8i:L, Fa:8u:L, Fa8il, Fu8aLah, Fu8uL, miF8a:L, miF8i:L, etc. See Wright (1967, vol.1, pp.133-40).

41 A. al- Akhḍar Ghazāl is the director general of the Institute of Researches and Studies of Arabization, Rabat, Morocco. The discussion referred to took place on 1 March 1980 during the author's fieldwork visit to the Moroccan capital.

3 Compounding : its role and significance

Introduction

The issue of compounding as a word formational process has given rise to much controversy among Arab writers and language reformers. Viewpoints have been widely different with regard to the significance of the role played by this process in the configuration of native vocabulary and whether or not it should be promoted as one of the viable means of generating scientific vocabulary in present-day Arabic. This chapter investigates the problem and the various factors underlying the linguistic, as well as extra-linguistic, attitudes involved.

It is obvious that Indo-European languages (mainly English and French) have been for the last two centuries the main sources from which Arabic draws its new scientific and technical terminology. The channel through which these terms are introduced into the language has been either arabization, i.e. borrowing, or rendition into native counterparts via a variety of methods which may or may not be in complete harmony with the typical character of the Arabic word. It is also true that the principle of compounding plays an increasingly important role in the configuration of the scientific vocabulary of the languages just referred to. In view of these facts, we have on frequent occasions found it necessary (or rather inevitable) to compare what is discussed with regard to Arabic with what is the case in other languages, especially English. This approach has gained us the advantage of being in a position to detect, identify and account for those phenomena of word formation or types of lexical structure that are not characteristic of Arabic, i.e. have evolved due to the impact of foreign influence.

The types of construction included range from single-word lexical units to set combinations and multi-word lexical expressions which are employed by Arabic to express what in European languages may be conveyed through single units based on the principle of compounding or affixation. We shall also discuss a number of attempts which reveal their authors' tendency to model their coinages on certain types of non-native formations. In a number of cases we have included examples from dialectal Arabic as well as others we have heard used by people from different walks of life. But prior to all this we have tried to trace the phenomena which are the subject of our analysis back to the early stages of Arabic.

3.1 The concept of Naḥt

More often than not Arab as well as non-Arab writers use the term "compounding" to refer to a word-formational process traditionally known in Arabic by the name of *an-naḥt*. This word, which is a derivative of the root n-ḥ-t "to chisel, hew (wood, stone, etc.)", has commonly been used by Arab grammarians to denote the principle of lexical creation whereby one or more radical consonants of more than one root take part in the formation of a single lexical item. The terms *manḥu:t* and *manḥu:tun minh*, respectively, denote the resultant form and the underlying elements involved in the process.

It should be noted, however, that in our analysis the term compound/compounding is not used to refer to the process just explained. This is due to the fact that naht-type words involve certain characteristics which may not be found in particular cases of compounding, as will soon be illustrated. Besides, Arabic has another, though related, type of process to which the term compounding may be more accurately applied (see 3.4).

A major characteristic of the naḥt-process is that words produced by it cannot be described as consisting of two roots (or root derivatives); rather they are one-root words resulting from an etymological process whereby two often semantically overlapping forms intersect due to their simultaneous occurrence over a long period of time. This fact is substantiated by a statement made about thirteen centuries ago by a celebrated Arab philologist, al-Khalīl bin Aḥmad (d.786), who observes that: "The Arabs, when they have overused two words, resort to naḥt, adding some of the letters of one of them to some of those of the other".(1) Another statement to the same effect is due to another ancient philologist, Ibn Fāris (d.1000), who tells us that "The Arabs coin [*tanḥat*] one word out of two others, which is a sort of abbreviation".(2) Instances of Arabic words illustrating the preceding statements are amply provided later; meanwhile it is useful to point out that lexical formations underlain by more or less the same process are also common in other languages. In English, for example, Jespersen (1922, pp.312-13) has identified the following instances:

blunt	blind + stunt
flush	flash + blush
glaze	glare + gaze (coined by Shakespeare)
slender	slight (slim) + tender
slide	slip + glide

There are also: brunch (breakfast + lunch); smog (smoke + fog); motel (motor + hotel). In French we find haut (Old French halt) from Germanic hoch + Latin altum.

The process yielding words such as the ones just cited may appropriately be referred to by such terms as "blending", "fusion" or "contamination".(3) The same is also applicable in the case of Arabic *manḥu:ta:t*, i.e. naḥt-type formations.(4) However, it seems useful that the latter be referred to by the Arabic term itself.

In the case of compounding, on the other hand, unlike the case of naḥt or blending, the elements constituting the compound words (e.g. doorhandle, headmaster, bedtime, etc.) are kept intact, i.e. they retain the same number of sounds they have at the pre-compound level. They are therefore more readily reatomized into their underlying constituent elements. An Arabic construction which may be described as the nearest in nature to this type is the one referred to by the term *al-murakkabu l-mazjiyy* "synthetic or mixed compound", an account of which is given in 3.4.

3.2 Naḥt in medieval Arabic

The question of naḥt was discussed by several medieval scholars among whom the aforementioned philologist, Ibn Fāris, stands out as the main contributor (1946, vol.1, p.328; 1963, p.134, fn.2). His contribution in this respect does not merely consist in attracting the attention to the existence of naḥt-featured words in the language, but also in his putting forward the view that most quadriliteral and quinquiliteral words in Arabic are products of this process. His definition (quoted above) constitutes the framework within which he presents his theory, which holds that a considerable number of Arabic words can be accounted for in terms of either of two processes, both of which are characterized by the feature of naht. On the one hand, there are words which (he argues) relate to two stems capable of independent existence in the language. (This fact may be verified by the independent occurrence of such stems in classical dictionaries and philological works.) For example: *ḍibaṭr* "strongly built (of a person)" from *ḍabaṭ-* "to hold fast, keep one's self-possession + *ḍabar-* "to be firmly made" (1963, p.271).

On the other hand, there are words which according to Ibn Fāris are accountable for in terms of addition (affixation). A triliteral word may have added to it one or more sounds which serve to modify its meaning, usually by introducing an element of intensiveness. For example *raᶜš* "shaking" + *n* *raᶜšan* "timorous" (1963, pp.99-100). We are also told that the added sound may be a prefix, an infix or a suffix. It is not made clear, however, where the added sound comes from, which in the light of modern linguistics proves the treatment to be inadequate. It is possible, on the other hand, to argue that the added sound is a remnant of an ancient form no longer operative, or which the writer was not able to establish. A pointer in this direction is the fact that a given word may at one time be accounted for in terms of naḥt proper, and at another in terms of affixation. A case in point is the word *ṣildim* "hard-hoofed",(5) of which two accounts are given, once as a naḥt-word from *ṣald* and *ṣadm* and once as deriving from the former with the sound *m* suffixed to it (Ibn Fāris, 1963, pp.271 and 100 respectively).

Mention must be made of the fact that early attempts to study naht were in general characterized by a considerable lack of accuracy and precision. To carry on with the account given by Ibn Fāris, we find that his claims can often easily be proved to be based on a purely impressionistic basis, thus yielding erroneous explanations. It seems that his enthusiasm for substantiating his theory and establishing naḥt as a major word-formational procedure in Arabic has even blindfolded him to the fact that what he considered to be a *manḥu:t* (i.e. a blend) may in actual fact be a non-Arabic word. A case in point is the strange account he gives of the word *jardab-* "to eat greedily, eat with one hand and push others back with the other" (1946, vol.1, p.506). He argues that the word derives from *jadab-* "to refuse to offer (food etc.)"(6) and *jira:b* "knapsack". The idea, he suggests, is that of a person who prevents others from reaching his food by protecting it with his hand which serves, as it were, as a sack in which the food is kept. However, had he been more adequately informed he would have discovered that *jardab-* is actually a verbal form derived from an old borrowing, namely *jurduba:n* from Persian girdaba:n "one who clasps his hands upon meat on a table, to prevent it from being carried off".(7) What is even more interesting is that the writer goes so far as to quote the latter in a verse, thus suggesting that it is an established Arabic word (1946, vol.1, p.506):

> *?iḏa: ma: kunta fi: qawmin šaha:wa:*
> *fa-la: tajᶜal šima:laka jurduba:na:*
> If you happen to be amidst covetous people
> let there be no jurduba:n to your left.

It may have already been realized that a borrowed word may be mistakenly accounted for in terms of naht due to the fortuitous similarities existing between what are thought to be its constituent elements and independent forms of native origin. We have been able to detect several instances of this phenomenon. To take one case (there are further examples in Chapter 4), another Persian loanform in Arabic is *farazdaq* "lumps of dough"; the original word is Per. bara:zda "a mass of leaven". An Arabic-speaking person without prior knowledge of the fact that *farazdaq* is an originally non-native word, if asked to analyse it into what may supposedly be its underlying constituents, would most likely give the two forms *faraz-* "to divide" and *daqq-* "to grind grain etc. so as to produce fine flour". Such an analysis would be prompted by the fact that these two forms exist in his linguistic system as semantically distinct linguistic units. And this is what seems to be the sort of rationalization on which Ibn Fāris has based his treatment of this loanform (1946, vol.4, p.513). He seems to have figured some semantic as well as phonological proximity between the loanform and the Arabic words, thus assuming that combined together they would imply the making of dough and its division into small pieces before it is baked.

However, the shortcomings revealed above need not be taken to imply the non-existence of naḥt in medieval Arabic. One should hasten to admit though that the method is nowhere as important in Arabic as that of analogical derivation (Chapter 2). It is also worth while to mention that naḥt-featured words are more often than not representative of cases where particular types of long constructions associated with certain conventional functions are rendered shorter for the sake of brevity. Below is a classification of these words, together with their underlying constituents.

3.2.1 Types of naḥt-words in medieval Arabic

(i) A new word of a quadriliteral or quinquiliteral root from two independent words based on triliteral roots. Examples:

ṣildim ṣ-l-d-m — hard-hoofed — *ṣald* < ṣ-l-d strong, hard; *ṣadm* < ṣ-d-m collision, striking (8)

bazmax- b-z-m-x — to be supercilious — *bazax-* < b-z-x to walk with the chest protruding and the back drawn in; *zamax-* < z-m-x to be haughty (9)

ṣahṣaliq s-h-s-l-q — strong, loud — *ṣahal* < ṣ-h-l rough or hoarse voice; *ṣalaq-* < ṣ-l-q to scream (10)

(ii) A new single word, mostly of a quadriliteral root from several elements which constitute its underlying construction. Examples:

ḥawlaq- ḥ-w-l-q la: ḥawla wa-la: quwwata ?illa: bi-lla:h
to say: there is neither might nor strength save in Allah (11)
dam^{c}az- d-m-^{c}-z ?ada:ma l-la:hu ^{c}izzak
to say: may Allah preserve your glory! (12)
basmal- b-s-m-l bi-smi l-la:hi r-raḥma:ni r-raḥi:m
to say: in the name of Allah, the Beneficent, the Merciful (13)

The type of formations exemplified above began to appear in Arabic with the advent of Islam, as can easily be inferred from the concepts expressed by the examples themselves. Unlike the previous category, these do not fall within the scope of etymological blending; rather they are conventional forms of abbreviated phrases and expressions which are regularly used as religious formulae. It remains to be added that new instances of similarly coined words are hardly produced nowadays.

(iii) A single relative adjective is formed from a bicomponental (sometimes multicomponental) proper noun, where the first noun is in the status constructus governing the second in the genitive:

^{c}abdariyy (a person) belonging to ^{c}Abd al-Da:r (the name of a family in Mecca)
^{c}abqasiyy (a person) belonging to ^{c}Abd al-Qays (name of a tribe)
^{c}abšamiyy (a person) belonging to ^{c}Abd al-Shams(name of a clan)(14)

The termination *-iyy* constitutes an important and a highly frequent suffix in Arabic, ancient or modern. It is used to indicate *an-nisbah* "reference or relation", and words that end in this suffix are known as *al-?asma:?u l-mansu:bah* "relative adjectives", which function both substantively and adjectivally, e.g. *^{c}Arab*; *^{c}Arabiyy* "an Arab; Arabic". This example is a characteristic instance of the application of this suffix, where the element it is added to is a single independent word. In the previous examples, however, a different process is involved. The element preceding the suffix this time is not a single independent word but rather a combination of parts of words representing the underlying constituents of the new (naḥt) form. An obvious explanation for the application of naḥt in these cases is that the underlying multi-componental construction does not lend itself easily for the derivation of a conveniently short nisbah-adjective.

It should be mentioned, however, that this procedure has not been applied consistently, nor has it been described by Arab grammarians as being the customary method of forming relative adjectives from constructions of the type under discussion, i.e. where the first component is in the status constructus, governing the second in the genitive, as in *da:ru l-ḥikmati* "the house of wisdom". What is described as the common and normal method in these cases, which indeed does not lack evidence from actual linguistic practice, can be summarized in the general procedure that one of the components of the construction is omitted (i.e. either the governing word or the one in the genitive) and the other one takes the suffix *-iyy* (see Wright, 1967, pp.149-65; Ḥasan, 1963, vol.4, pp.556-8). Which one of the two is omitted is determined by whether or not it represents the more important element in indicating the person, object or concept referred to. Thus are obtained:

ḥanafiyy	abu: ḥani:fah, the founder of an orthodox school of theology
?anfiyy	?anfu n-na:qati, (camel's nose) nickname of a man after whom his sub-tribe was named
ḥija:riyy	wa:di: l-ḥija:rati, Guadalajara in Spain
taymiyy	taymu l-la:ti, name of a clan
mana:fiyy	ᶜabdu mana:fin, name of an ancient clan

Despite the unfavourable attitude generally held by traditional grammarians towards formations of the earlier type, it is not impossible to find examples coined recently and by prominent Arab scholars. Ibrāhīm Anīs, for instance, uses the term *darᶜamiyy* as a nisbah-adjective from *da:ru l-ᶜulu:m* "the house of sciences" (the name of a college in Cairo)(Anīs, 1958, p.76). Another example is *?anfamiyy*, coined by the same writer to denote the sound produced through the *?anf* "nose" and *fam* "mouth" (Anīs, 1950, p.68). And we wonder if it would be acceptable to derive *rasxamiyy* from *ra?su l-xaymah*, the name of an Arab emirate.

3.2.2 General notes and conclusions

In the light of the above statements, it is possible to draw a number of conclusions concerning the process of naht in medieval Arabic and its role in the configuration of the native vocabulary. We have already mentioned the fact that the process in question is far less productive than that of derivation through ?ištiqa:q. Unlike ?ištiqa:q, there has been no direct statement by Arab philologists, not even by Ibn Fāris himself, as to the acceptability or analogicality of naḥt as a productive model on which new formations could be patterned. Nor is there any mention of the way it works; apart from the statement that it consists of the formation of one word from two others, there are hardly any rules concerning the morphological segmentation of the underlying components or the order of the manḥu:t-constituting elements.(15)

The rarity or low frequency in Arabic of words or parts of words merged together may be accounted for in terms of certain inherent characteristics of its structure and its psychological attitude towards linguistic material. It is a deeply ingrained tradition in the heart of the Arabic language that a word be organically related to an origin or root whose basic (consonantal) structure it bears in all its occurrences, and is the property it shares with other words in the same paradigm (for examples see 1.3.2). Unlike the case in many European languages, in Arabic the free word is the most frequent lexical unit.(16) In other languages a word may be composed of more than one root morpheme, hence the term "polymorphemic" unit or word. Two or more roots may occur in the same word, with or without affixes. Such combinations are usually labelled "compound words", e.g. doorknob (root + root) and caretaker (root + root + affix). A compound word may also contain one or more bound forms, e.g. "electrostatic" (electro- + stat- + ic). Compared with other languages, English however seems to be more in favour of word groups than long compound words. German, on the other hand, is particularly distinguished for its long compound words where several roots and affixes may be used. Take for instance the word "Kriegsverpflegungsamt", commissariat department, (Krieg-s-ver-pfleg-ung-s-amt: root + affix + affix + root + affix + affix + root) (see Robins, 1975, p.201).

Mentioning English and German, it is relevant to point out the psychological contrast between these two languages, as illustrated by Sapir. English, as he puts it, "has long been striving for the completely unified, unanalysed word, regardless of whether it is monosyllabic or polysyllabic". In German, on the other hand, "polysyllabic words strive to analyse themselves into significant elements". Thus in English a word like "credible" is entirely welcome since it represents a unitary idea and does not present the unconscious mind with the necessity of analysing it formally, i.e. "cred-" does not have the same position as that of "mad-" in madness. In German, however, the Latin-German "kredibel", borrowed at the height of certain cultural influences, soon dropped out of use because it was not analysable into significant elements, and so ran counter to the customary method of word-formation and linguistic feeling. (See Sapir, 1921, pp.195-6.)

It may thus be said that the opposition of a particular language to certain modes of lexical creation is largely due to the incongruity of the principles underlying those modes with the characteristic nature of its structure. As far as Arabic is concerned, compounding as exhibited by European languages is not one of its propensities, and Henri Fleisch is right in suggesting that "L'arabe ne peut réunir deux mots par une voyelle thématique comme font le latin et le grec, ni les joindre selon les composés de l'anglais ou de l'allemand. La composition n'est pas dans son génie." (Fleisch, 1956, p.124). What may be said to represent the natural tendency of the langauge is the application of the root-and-pattern system in which these two components interrelate in such a way as to give the process a uniquely distinctive character. These interrelations are nicely depicted by Cantineau (1950, p.74) in the following passage:

> Chaque mot a sa racine et son schème; on pourrait comparer le vocabulaire à un tissu dont la trame serait l'ensemble des racines attestées dans la langue, et la chaîne l'ensemble des schèmes existant. Chaque point d'intersection de la chaîne et de la trame serait un mot, car tout mot est entièrement défini sans ambiguité par sa racine et son schème, tout schème de son côté fournissant des mots à différentes racines, et la plupart des racines fournissant des mots de différents schemes.

Having formed an idea of the role of naht in medieval times, let us now turn to the state of affairs in the present circumstances. Before we embark upon this, it is worth mentioning that modern attempts at adopting this process have been largely confined to the language of science. This may be accounted for by the nature of modern scientific vocabulary in general, the impact of foreign terminology and the urgent need for new vocabulary to cope with the ever-increasing demands of modern civilization.

3.3 Naḥt in modern Standard Arabic

In this discussion we shall continue to use the term "naḥt" for those cases where one or more of the underlying constituents are partially represented in the resultant form.

3.3.1 The position of Arab language academies

The views held by Arab language academies regarding naht are largely in agreement with those of medieval grammarians as expounded in the previous discussions. Ibn Fāris's definition quoted above is almost literally echoed by the Cairene Academy: "Naḥt is a sort of abbreviation; it consists in the formation of one word out of two or more others".(17) The academy also specifies a number of features as being characteristic of this process. These include:

(a) the underlying constituents of the manḥu:t need not necessarily all be represented in it, e.g. *al-kabtaᶜah* from *kabata l-la:hu ᶜaduwwak* "May God put your enemy to shame!";
(b) the first word in the underlying construction need not be retained in its original shape in the manhu:t, which is almost exclusively the case in the examples we have seen; and
(c) the short vowels (*ḥaraka:t*) and zero-vowels (*sakana:t*) of the letters constituting the underlying elements need not be observed in the manhu:t. For instance, in *maškanah* *ma: ša:ʔa l-la:hu ka:n* "whatever is intended by God happens", the *š* is unvowelled, unlike the case in the underlying word containing it (see al-Ḥuṣriyy, 1958, pp.140-1).

It is thus obvious that the features specified by the academy are fundamentally general statements already implied, explicitly or implicitly, in the defining or treatment of naḥt proposed by medieval grammarians. What may be described as a more significant contribution or development, however, is the academy's authorization of this procedure in scientific terminology. This authorization comes in a statement issued by a committee composed of a number of academicians especially appointed for assessing the role and potential of naḥt and the possibility of adopting it as a further means of lexical expansion. The statement reads: "We [the committee] agree to the legality of naḥt in scientific and technical disciplines due to the urgent need to express their concepts in concise Arabic terms." However, following several discussions in the academy, this statement was subjected to the following restriction: "Naḥt is permissible [only] when necessitated by scientific need", which, due to its vagueness, gave rise to much controversy.(18)

The phrase "scientific need" has been interpreted differently by different scholars according to the type of orientation and predilection they have. Thus, those who are in favour of naḥt have taken it to be the long-awaited go-ahead, and therefore set out to apply the principle in the most generous terms. On the other hand, those with a more conservative attitude have understood it to be of a rather restrictive force, permitting naḥt only in very rare cases. In addition to these two categories, there are also those whose attempts are characterized by a strong foreign influence embodied in their mapping of new Arabic formations on foreign patterns (see 3.3.2 and 3.3.4).

In compliance with these rulings and principles of the Cairene Academy, it has sanctioned a number of naḥt-featured words which may be represented by the following sample:

šibza:l	*šibhu z-zula:l*	albuminoid
ḥalkaḥ-) *ḥalkal-)*	*ḥallala l-kuḥu:l*	alcoholyse
šibqily	*šibhu qily*	alkaloid
faḥma?iyya:t	*faḥm + ma:? + -iyya:t*	carbohydrates
la?man;	*la: + ?ami:n*	deaminate (19)
laklar-	*la: + klo:r*	dechlorinate (19)
lama?-	*la: + ma:?*	dehydrate
ḥalma?-	*ḥallala bi'l-ma:?*	hydrolyse (20)

The form *la:ma:?iyy* "anhydrous" is also included in the original list as an example of a naḥt-produced word. From the point of view of our analysis, however, such a form is not included in the category of naḥt since it lacks the criterion of shortness, i.e. none of its underlying constituents (*la: + ma:? + -iyy*) is partially represented in it. This distinction is significant in that, among other things, it correlates with another one (viz. understandability) on which the acceptance or rejection of a new form depends (see 3.3.4 - 3.4).

The Cairene Academy also stipulates that "naḥt-words are supposed to keep within the limits of comprehensibility" (p.158). In other words, the new forms should not strike the native hearer/reader as entirely new and unintelligible, the constituent elements should not be so shortened or modified as to make it difficult for one to find some connection between them and their underlying constituents. It is on this basis that such coinages are considered acceptable as:

kahramağnaṭi:siyy	*kahraba:?iyy + mağnaṭi:siyy*	electromagnetic
kahraḍaw?iyy	*kahraba:?iyy + ḍaw?iyy*	photoelectric
šibğarawiyy	*šibhu ğarawiyy*	colloidal

Whereas others, such as the following, are rejected on the grounds of incomprehensibility (see al-Khaṭīb, 1978, p.741, fn.1):

nazwarah	*naz^cu l-waraq*	defoliation
ḥarṣam-	*ḥarrara mina ṣ-ṣamġ*	to degum
zahraj-	*?aza:la l-hi:dro:ji:n*	dehydrogenate

More often than not, both in language academies and other concerned circles, another criterion put forward for judging the practicality and acceptability of a new coinage is its conformity to what is described as "the Arabic taste" (*ad-ḏawqu l-^carabiyy*), an all-embracing concept in which a large variety of considerations pertaining to the typical character of the Arabic word may be said to be embodied. For instance, al-Shihābiyy, a member of the Syrian Academy, stipulates that he who intends to employ naht must necessarily be capable of sensing whether the new form is compatible with the Arabic taste or not (see al-Shihābiyy, 1959, p.548; 1955, pp.98-101).

It must be admitted that the criterion just mentioned, as well as the earlier one concerning "scientific need", is too general and too vague to be of any practical value. It is not always within our capability, especially in the language of science, to shape words in such a way as to make them agreeable to our taste. The nature of the factors involved in the creation of a new term may be such that it is not always possible for us to enjoy this facility. Besides, the concept of taste is often indeterminable, and two persons may have varied opinions regarding one and the same word. Moreover, we have already cited cases where the incorporation of foreign words or parts of words is seen as necessary due to the fact that, for one reason or another, adequate native equivalents in these cases are not readily available. And if these foreign elements were not possibly susceptible to complete assimilation – as we shall see in Chapter 4 – what would be done to make them conform to the present criterion?

It is not clear either what constitutes a "scientific need". A term used by a scientist for a particular object, concept or phenomenon may, under certain circumstances, be of equal importance to any person from any walk of life. Despite the fact that certain features can be distinguished as characteristic of scientific vocabulary, as opposed to literary or ordinary vocabulary (4.4.3.2), it is not (at least not always) possible to make a clear-cut distinction between scientific and non-scientific needs.

Regarding the criterion of understandability, it goes without saying that the more transparent the new coinage, the more useful and practical it is. For example, the Arabic rendition of "gastrostomy", namely *fatḥu l-mi^cdah* would be more readily understood than the naḥt-form *fatmi^cdah*; the latter in turn is more understandable than the yet shorter manḥu:t *fat^cadah* (this coinage is from Jirjis; see others proposed by him on page 70).

In brief, it could be stated in general terms that Arab language academies have been conservative with regard to the employment of naḥt as a word-formational procedure in Arabic. It is now time we turned to the situation outside the context of the academies.

3.3.2 Modern Arab scholars and the role of naḥt: views and attempts

The opinions put forward by many Arab scholars regarding the practical value of naḥt are extremely varied and sharply divided. It is therefore useful to give an illustrative account of these opinions highlighting the most important points.

It was mentioned earlier that some scholars adopt a conservative view towards the adoption of naḥt in SA. Their main argument for this claims that the process is not characteristic of Semitic languages in general and Arabic in particular, and that words produced by this method are very few in these languages and often denounced as vague and abstruse. Such views are expressed by ᶜAli ᶜAbd al-Wāḥid Wāfi (1972, pp.187-9; he argues that "the Arabic words that are formed from two or more independent roots do not exceed some tens [in number]"); Anastās al-Karmaliyy (1928, vi, p.293); Aḥmad ᶜAli al-Iskandariyy (1935, p.17); Muṣṭafa Jawād (1955, p.86: "talking of naḥt, I would like to point out that I do not rely on it in the new expressions because it is rare in Arabic and distorts its vocabulary); Maḥmūd Fahmi Ḥijāziyy (1978, p.93), etc.

The attack on naḥt is also frequently backed by an emphasis on the merits of the system of ?ištiqa:q. Discussing the subject with Ismāᶜīl Maḍhhar, al-Iskandariyy remarks that Arabic may have employed naḥt in its early developmental stages, but now it is no longer necessary, "its door is closed", because the language has adapted itself and become a language of ?ištiqa:q, not of naḥt (see Maḍhhar, "Tajdīd...", n.d., p.17).

The resistance with which new attempts to employ naḥt are sometimes confronted is best illustrated by the following situation. Mārūn Ghuṣn, an enthusiastic supporter of naḥt, proposed a number of instances where the present procedure is applied (1935, p.302):

?arbayad	*?arbaᶜ* "four" + *yad* "hand"	quadrumane
?arbarijl	*?arbaᶜ* "four" + *rijl* "foot"	quadrupède
ḏu:ṯad	*ḏu:* "possessing" + *ṯady* "breast"	mammifère

By way of commenting on the above constructions, Salīm al-Jundi (1935, p.361), an academician, had this to say: "I swear by God that had I before heard someone say *?arbayad*, *?arbarijl*, or *ḏu:ṯad* ... I would not have doubted their being a jesting buffoon or a feverish person raving away". Another writer, Saᶜīd al-Afghāniyy (1936, p.152), wrote an article entitled "A Neutral View" (*Kalimatu Ḥiya:d*) whereby he tried to settle the controversy between Ghuṣn and his opponent. However, he himself proved to be no less critical of Ghuṣn than al-Jundi; the statement concluding his article reads: "What Mr Ghuṣn is advocating is death [i.e. to Arabic] beyond all doubt".

There are, on the other hand, a considerable number of writers and dictionary compilers who exhibit a remarkable degree of interest in naht and have made some attempts to put their views into practice despite the stubborn resistance on the part of academicians and purists. According to these writers, naḥt is a useful means by which the terminological deficiency of modern scientific Arabic can be overcome, and that traditional methods alone do not solve the problem.

Thus Jirjis, for one, has proposed the employment of parts of certain Arabic words as combining forms in medical terminology:

qaṭ-	*qaṭ*^{c}	cutting
ṣal-	*?isti?ṣa:l*	removal
fat-	*fatḥ*	opening
waj-	*waja*^{c}	pain

These elements are combined initially with others taken from other words. As will be seen from the examples below, the resultant forms of such combinations reveal close correspondence to certain patterns of word formation adopted by foreign languages. Here is a sample of the words coined by Jirjis (1961, p.66):

qaṭjarah	< *qaṭ-* + *-jarah* < *ḥanjarah*	"larynx"	laryngotomy
ṣalkalah	< *ṣal-* + *-kalah* < *kulyah*	"kidney"	nephrectomy
fatkalah	< *fat-* + *-kalah* < *kulyah*	"kidney"	nephrostomy
waj^{c}*adah*	< *waj-* + *-*^{c}*adah* < *ma*^{c}*idah*	"stomach"	gastralgia

Another perhaps more radical attempt is that made by Ismā^{c}īl Maḍhhar ("Tajdīd", n.d., pp.25-6), who tells us that after long experience as a specialist in zoology and botany, he has come to the conclusion that the need for new technical terms in the present age is too great and too urgent to be met by analogical derivation only. He therefore opens the door for naht as an indispensable means of lexical expansion. An example of his attempt may be represented by his rendition of the Greek word "hypsiprymnodontinae" into Arabic *al-*^{c}*awsaniyya:t*. The former is reatomized into its constituent components which are then represented by Arabic elements taken from a number of words which otherwise constitute a multiword lexical unit equivalent in signification to the Greek word. The process may be illustrated as follows:

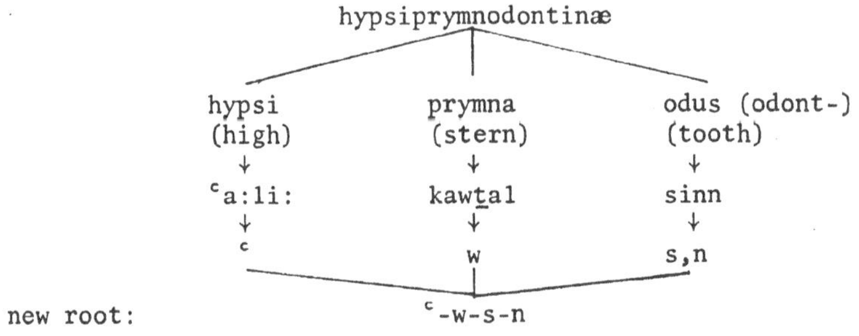

From this new root and using the pattern Fa8LaL, he thus derives the new terms ^{c}*awsan*, any of the class of animals that have teeth with high crowns, ^{c}*awsaniyy*, a relative (nisbah) adjective, and ^{c}*awsaniyya:t*, the class or species of animals so specified (p.73).

Support for naḥt comes also from Sāṭiᶜ al-Ḥuṣriyy (1958, pp.130-47), a distinguished Arab scholar. According to him, naḥt is particularly advantageous in that it produces economical terms and thus helps us avoid the use of long and varied constructions, the latter being unfavourable from the point of view of scientific language. Thus he prefers type (a) to type (b) of the following instances:

centrifugal force	(a)	*al-quwwatu l-ᶜamarkaziyyah*
	(b)	*al-quwwatu ṭ-ṭa:ridatu ᶜani l-markaz*
somnambulism	(a)	*as-sarmanah*
	(b)	*as-sayru fi: l-mana:m*
prehistory	(a)	*al-qabtaᶜri:x*
	(b)	*(ma:) qabla t-taʔri:x*

Instances of naḥt constructions also occur, sometimes abundantly, in a number of bilingual (English/French-Arabic) dictionaries, whether general or specialized. Here again the size and nature of the phenomenon varies from one dictionary to another. For instance, al-Baᶜlabakkiyy (1973), al-Kawākibiyy (1947) and Maḍhhar ("al-Nahḍah", n.d.) use a relatively large number of naḥt-featured words; others, e.g. al-Khaṭīb (1978, p.119, fn.4), do not exhibit a similar degree of enthusiasm for such formations.

3.3.3 The problem in the light of modern linguistic theory

The state of affairs just discussed reveals the fact that there is a sharp division among Arab writers with regard to the application of the naht process and the extent of its conformity to the native system. Most of the views we have examined are characterized by extremity and inflexibility; a lack of a clear understanding of language and linguistic laws is also discernible. For instance, the fact that ?ištiqa:q represents the characteristic method of word derivation in the language seems to have been employed by some writers as an argument against any form of development or growth outside the realm of this method. This is an extreme view, for any language is always susceptible to adaptation in accordance with the type and nature of the society it serves, and the Arab society, like any other, is in an everlasting process of change and development. Scientific Arabic nowadays is confronted with a need for new terminology such as it has never experienced before. It should not be surprising then that it resorts to some form of adaptation whereby it may fill what seems to be a serious gap, while at the same time it makes full use of its characteristic system.(21) It should be understood, however, that we are not implying that the door should be kept open to all sorts of attempts without taking into account certain considerations.

On the other hand, the attempt is sometimes carried too far. We have seen cases where some scholars apply to Arabic what may at first sight seem to be a similar procedure to that adopted by other languages in which mainly Greek and Latin roots are commonly used in the construction of scientific and technical vocabulary. Maḍhhar's *ᶜawsaniyya:t* (p.70) is one case in point. However, a closer examination of the situation in question reveals a number of factors which seem to have escaped the attention of the authors of such attempts. One of these factors is that these Greek and Latin elements (and this is not the case in Arabic) are easily distinguished within the word by virtue of the fact that they represent semantically distinct elements capable of specific lexical meanings, even though they do not occur as free words. Besides, in languages where the combination of such elements is a common practice or widely applied, the speakers, depending on their linguistic experience, are capable of recognizing the meaning of the whole word by virtue of their prior knowledge of the meaning of the individual elements that constitute it. The combining of two or more of such elements does not affect or obscure the character of any of them, which accounts for their capability to occur in different parts of the word, i.e. initially, medially and/or finally. Take for instance the combining form "anthrop-" (Gr. anthrōpos) which may occur initially, as in "anthropology" or finally as in "misanthrope". Compare also the element "morph" (Gr. morphē "form") in "morphallaxis", "anthropomorphotheist" and "pseudomorph".

From the point of view of Arabic, however, the state of affairs is hardly the same. Such mobility of the elements constituting the word, as illustrated above, is not allowed. We have mentioned in our account of Arabic word structure that one or more elements may be added to a simple root-derivative to form a new complex form, a process whereby the relationship between the latter and the basic idea of the root is still maintained, though with some modification. However, the added elements are always derivational or inflectional affixed which, together with the root morpheme, form fixed measures. The positions occupied by these affixes are invariably fixed, i.e. with respect to other constituents of the word. In addition, as has already been hinted, no matter how complex an Arabic word may be, the main idea is always a property of the root; other elements do no more than cluster around performing a subsidiary function. The same is not true in the case of "anthropomorphotheist", for instance, where more than one element may be said to represent a main idea.

An important point to be stressed here is that the Arabic speaker has been brought up with the linguistic habit of constructing words as free one-root units, not as combinations of semantically distinct elements, unless these are multi-word lexical constructions, or set combinations. The root constitutes an essential component of the word and is always present in the speaker's subconscious. It follows then that according to his intuitive knowledge of his language, the Arabic speaker cannot be expected to understand a form like *faṭkalah* (p.70) as representing a more-than one-root word; rather he would take it to be a derivative of the single root *f-t-k-l, which does not exist in the language. The fact that the form in question corresponds to *fat-* <*fatḥ* "opening" + *-kalah* <*kulyah* "kidney" is of no significance because these abbreviated elements are not capable of representing their underlying forms, and consequently have no semantically distinct character. *fat-*, for instance, may be taken to relate to any root-derivative where the first and second radicals are *f* and *t* respectively, e.g. *fatar-* "to subside", *fataq-* "to undo the sewing", *fatal-* "to twist", etc.

In view of these facts, it may safely be argued that coinages of the type represented by *ᶜawsaniyya:t* and *fatkalah* contradict the fundamental nature of Arabic and are therefore unlikely to achieve any success in actual linguistic practice (see Chapter 5).

Another important point to be raised here is that resort to naht sometimes seems to be motivated by no other reason than that the foreign counterpart is a compound word; otherwise what is wrong, for instance, with *fatḥu l-kulyah* vis-à-vis "nephrostomy"? The fact that in a certain language something is designated by a single word — though this in fact is often compounded of more than one constituent, as in the preceding example — need not necessarily mean that its equivalent in another language should be a single word too. Languages vary a great deal in this respect. We have seen cases where Arabic uses more-than-one-word lexical units for single (compound) English counterparts, and examples to the contrary are not at all difficult to provide:

before	→	a little before	=	*qabl*	→	*qubayl*
after	→	a little after	=	*baᶜd*	→	*buᶜayd*
above	→	a little above	=	*fawq*	→	*fuwayq*
below	→	a little below	=	*du:n*	→	*duwayn*

In addition to the instances of naḥt-words included in the preceding discussions, present-day scientific Arabic reveals a number of other types of constructions based on the same principle (naḥt) and which may be distinguished from one another on the basis of certain characteristics. Section 3.3.4 provides a classification of these types as well as an account of their potential and practicability in the language. Before we start this discussion, however, we need a brief account of the term "set combination" which will occur frequently in the following pages.

Set combinations

We have frequently used the term "multiword lexical unit" to designate those combinations of words which are capable of functioning as single wholes and may therefore be said to convey a lexical meaning. In other words, as far as their designation of segments of the extra-linguistic world is concerned, these combinations are on a par with lexical units consisting of one word only. They have the characteristics that they are not created by the speaker on the spur of the moment, they do not vary with the circumstances, and their frequency of occurrence is not limited. In this way, these more-than-one-word single units are set apart from what may be described as "free combinations" which we create ad hoc to suit the purpose which we happen to want to fulfil, and which, by virtue of the fact that they derive their meanings from that of single combined words, cannot be said to function as wholes or convey a lexical meaning. A useful and economical term to designate the first type of combination, i.e. more-than-one-word one unit, is "set combination" (Zgusta, 1971, pp.140f.), which is applicable to such designative multiword Arabic units as:

ka:timu ṣ-ṣawt	silencer (lit. sound silencer or suppressor)
šibhu jazi:rah	peninsula (lit. quasi-island)
na:ṭiḥa:tu s-saḥa:b	skyscrapers (lit. cloud gorers)

However, in view of the fact that free combinations are not always easily distinguishable from set combinations, we shall employ the latter in a broad sense to refer to various types of multiword lexical constructions which Arabic uses vis-à-vis foreign compound or complex words.

It is worth while to mention a number of features which we find characteristic of Arabic set combinations:

(a) As far as scientific Arabic is concerned, set combinations in the majority of cases result from the reatomization of compound or complex words into their native constituent elements which are then translated into native counterparts. Examples:

zero gravity	*at-ṯuqlu ṣ-ṣifr*
antimissile	*qaḏi:fatun muḍa:ddah*
supersonic	*fawqa s-samᶜiyy* (acoustics) *fawqa ṣ-ṣawtiyy* (aeronautics)

(b) It is frequently the case that an Arabic set combination has a native synonym consisting of one word only, which is usually produced by means of ?ištiqa:q, e.g.

miqya:su l-ḥara:rah	=	*miḥra:r*	thermometer
ġayru mustaqirr	=	*qaliq*	unstable
ᶜadamu l-?inḥiya:z	=	*al-ḥiya:d*	nonalignment

(c) The overall meaning of a set combination may not necessarily be literally representative of that of the individual constituents:

al-ḥarbu l-ba:ridah	the cold war
al-?iḏa:ᶜatu l-mar?iyyah	television (lit. the seen broadcast)
al-qamaru ṣ-ṣina:ᶜiyy	sputnik (lit. the artificial moon)

(d) The relation of a simple to a compound or complex word in English is often paralleled in Arabic by one between a one-word unit and a set combination. For example:

action	↔	*fiᶜl*
reaction	↔	*raddu l-fiᶜl*
analysis	↔	*taḥli:l*
psychoanalysis	↔	*taḥli:lun nafsiyy*
to saturate	↔	*šabbaᶜ-*
saturable	↔	*qa:bilun lit-tašabbuᶜ*
unsaturable	↔	*ġayru qa:bilin lit-tašabbuᶜ*

(e) The number of words constituting an Arabic set combination is, as may have already been gathered, often equivalent to that of the constituents of the foreign counterpart.

3.3.4 Types of naḥt-construction in modern scientific Arabic

It was pointed out in 3.3.3 that naḥt-formations like *^{c}awsaniyya:t* and *fatkalah* are not likely to survive or gain acceptance due to the fact that their meanings are difficult to derive from their constituent elements. From the point of view of understandability, such formations may thus be referred to by the term "Opaque N-constructions" (N standing for naḥt). Regarding their morphological structure, they are characterized by the fact that none of their underlying constituents is kept intact; all are subjected to the process of abbreviation. From this point of view then we may refer to these formations as "Full N-constructions". The following instances, coined by a number of Arab writers and lexicographers, are representative of this category (set combinations are given in parentheses): (further examples are given in the appendix)

ḥalqaḍah(22) daydream	<	*ḥulm* "dream" + *yaqḍah* "waking" *(hulmu l-yaqdah)*
as-sarnamah(23) somnambulism	<	*as-sayr* "walking" + *an-nawm* "sleep" *(as-sayru xila:la n-nawm)*
al-li+nafiyya:t malacopterygii	<	*layyin* "soft" + *zi^{c}nafah* "fin" *(layyina:tu z-za^{c}a:nif)*
al-baṭjal(24) gastropod	<	*baṭn* "abdomen" + *rijl* "foot" *(al-baṭniyyu l-?arjul ?aqda:m)*
an-nazjanah(25) dehydrogenation	<	*naz^c* "elimination" + *al-hi:dro:ji:n* "hydrogen" *(naz^{c}u l-hi:dro:ji:n)*
az-zahrajah(26) dehydrogenation	<	*?iza:lah* "removal" + *al-hi:dro:ji:n* "hydrogen" *(?iza:latu l-hi:droji:n)*
tama:kub(27) Fr. isomérie	<	*tama:ṯul* "agreement" + *tarakkub* "composition" *(at-tama:ṯulu bit-tarki:b)*
ḥayšanah(28) Fr. bryozoaire	<	*ḥayawa:n* "animal" + *?ašanah* "moss" *(hayawa:nun tuhlubiyy)*
kahra:ṭi:s(29) electromagnet	<	*kahraba:?* "electricity" + *maġnaṭi:s* "magnet" *(al-maġnaṭi:su l-kahraba:?iyy)*
al-ḥarkabah(30) electromotion	<	*al-ḥarakah* "motion" + *al-kahraba:?* "electricity" *(al-ḥarakatu l-kahraba:?iyyah)*

We now turn to another type, "Semitransparent N-constructions", in which the combining of the different segments seems to follow certain regular patterns. In each case one of the underlying constituents, which is almost always the final one, retains its full shape in the construction, whereas the other is only partially represented. From the point of view of their morphological structure, such cases may therefore be designated by the term "Partial N-construction". We have also noticed that certain significant distinctions may be made within the present type. There are thus (a) those in which the abbreviated constituent is an "Open-class item" and (b) those where it is a "Closed-class item". Let us illustrate these statements starting with subtype (a).

The coinage *musjana:ḥiyya:t* (31) "orthoptera" consists of the element *mus-* which is an abbreviated form of the underlying constituent *mustaqi:m* "straight" and the fully represented component *jana:ḥ* "wing", plus the species-indicating suffix *-iyya:t* which is not relevant to the present discussion. As has already been implied, the coinage cannot be said to be fully intelligible to the speaker who has had no prior acquaintance with it. It would be quite easy for him to isolate and realize what is meant by *-jana:ḥ* by virtue of the fact that it represents part of his system; but what about *mus-*? Instead of relating it to *mustaqi:m*, he might understand it to be derived from *mustaṭi:l* "long", *mustatir* "concealed", *mustaqwis* "formed like a bow (*qaws*)", etc., as all these adjectives may be used as qualifiers of the noun *jana:ḥ*. Note also that the abbreviation *mus-* does not constitute part of the root underlying the word *mustaqi:m*, i.e. q-w-m, whereas it has been illustrated above that the root is an essential component of the Arabic word and represents the main idea underlying it. Being void of any semantic specification, this abbreviation is therefore very likely to be given a different interpretation than that intended by the coiner or neologizer. It is interesting to note that the coiner of *musjana:ḥiyya:t* himself has done so, even though the element he has employed this time *does* constitute part of the root. (32) He has rendered "electromotive" into *ḥarkabiyy*, taking *ḥar-* from *ḥarakiyy* "motive" ḥ-r-k, and *-kabiyy* from *kahrabiyy* (an alternative form of *kahraba:?iyy*) "electro-" k-h-r-b. However, he uses the same construction, *ḥarkabiyy*, for another term: "thermo-electric", where this time the abbreviation *har-* does not belong to *ḥarakiyy* but rather to *ḥara:riyy* "thermo-", where the root is ḥ-r-r and not ḥ-r-k (Maḏhhar, "al-Nahḍah", vol.i, p.461; 2, p.2241).

Here are some more examples of Partial N-construction where the abbreviated constituent is an open-class item:

kahra:kid electrostatic	*kahraba:?* "electricity" + *ra:kid* "static" *(kahraba:?iyyun ra:kid)*
jaḏra?siyya:t rhizocephala	*jaḏr* "root" + *ra?s* "head" *(jaḏriyya:tu l-?ar?us)*
kahraṣalbiyy Fr.electro-negatif	*kahraba:?* "electricity" + *ṣalbiyy* "negative" *(sa:libu š-šaḥnati l-kahraba:?iyyah)*
?arbayad quadrumane	*?arba*ᶜ "four" + *yad* "hand" *(ruba:ᶜiyyu l-?aydi:*

nafjismiyy "psychophysical" *nafs* "soul,psyche" + *jism* "body"

ᶜafnaba:t "saprophyte" *ᶜafan* "decayed organic matter"+*naba:t* "plant"

al-katṣu:f "linsey-woolsey" *al-katta:n* "linen"+*aṣ-ṣu:f* "wool"

With regard to subtype (b) of Partial N-construction, the abbreviated constituents are closed-class items. In view of the fact that the number of items constituting a closed class is fixed and relatively limited as opposed to that of an open one, and that the members of the former occur frequently in the language, it is possible to argue that naht-constructions of the present type are more easily distinguished and hence more likely to be understood than those of subtype (a). This being the case, Arabic separable prepositions, which are mostly triliteral, may be reduced to biliteral units and employed as prefixes, a process which has a great potential for meeting the urgent need for economical scientific terminology. Examples illustrating this statement will soon be provided; meanwhile here is a list of Arabic prepositions which may be used as prefixes in accordance with the above method:

qab-	<	*qabla*	before, pre-
gib-	<	*gibba*	after, post-
faw-	<	*fawqa*	above, over, super-
taḥ-	<	*taḥta*	under, beneath, sub-
du:-	<	*du:na*	below, sub-
bay-	<	*bayna*	between, among, inter-
ḥaw-	<	*ḥawla*	around, about, circum-
xal-	<	*xalfa*	behind, post-
ḍim-	<	*ḍimna*	in, within, inside of, among, intra-

To illustrate the application of these prefixes, as well as the statement preceding them, here are a number of examples where the use of naht-featured words in (a) may be compared with the alternative counterparts which are produced by traditional methods in (b):

(a) *al-?aᶜra:ḍu l-qabsulliyyah*
pretuberculous symptoms
(b) *?aᶜra:ḍu ma: qabla s-sull* or *al-?aᶜra:ḍu s-sa:biqatu li-ḏ̣uhu:ri s-sull*

(a) *an-naba:ta:tu t-taḥbaḥriyyah*
submarine plants
(b) *an-naba:ta:tu n-na:miyatu taḥta saṭḥi l-baḥr*

(a) *takawwuna:tun gibjali:diyyah*
postglacial formations
(b) *takawwuna:tu ma: baᶜda l-ᶜaṣri j-jali:diyy*

(a) *ṭa:?iratun fawṣawtiyyah*
supersonic plane
(b) *ṭa:?iratun fawqa ṣ-ṣawtiyyah* or *ṭa:?iratun ?asraᶜu mina ṣ-ṣawt*

(a) *faḍa:?un baynajmiyy*
interstellar space
(b) *faḍa:?u ma: bayna n-nuju:m* (for more examples see appendix)

An important argument in favour of accepting the utilization of Arabic prepositions in the way just demonstrated is that the process involved does not constitute an altogether novel phenomenon in the language. Mention has been made of the fact that Arabic contains a number of prepositions and particles which are always united with the following noun, the only difference being that these are sequences of one, rather than two, consonants, plus a short vowel. Consider the following examples:

li- as in *li-ḏa:lika* "for that reason"
bi- as in *bi-n-naha:r* "during the daytime"
ta- as in *ta-lla:hi* "by God!"

fa- as in *yawman fa-yawman*	"day after (by) day"
ka- as in *ʔana: ka-ṭa:libin*	"I as a student"

Hybrid naḥt-constructions

As far as the origin of their underlying constituents is concerned, the types of N-constructions we have so far discussed may be generally referred to by the term "Native N-constructions" since the constituents underlying them are of Arabic origin. There are however formations where an element of non-native origin is involved in the process. The term "Hybrid N-constructions" may be used to designate such formations. This phenomenon is largely restricted to chemical terminology, where the foreign element, a suffix or combining form, is attached to a full native word. The following examples are from al-Kawākibiyy (1947):

ḥamḍ-i:l	acyle	(p.6)
ᶜiṭr-i:l	aryle	(p.13)
xamr-i:l	vinyl	(p.69)
faḥm-i:d	carbonide	(p.17)
sukkar-i:d	glucoside	(p.34)
maᶜdan-i:d	metalloid	(p.43)
bawl-i:d	ureide	(p.69)
ṣu:f-i:n	lanolin	(p.41)

There are also cases where a native stem appears with several borrowed terminations. A case in point is the word:

kibri:t Eng. sulphur,	Fr. soufr:
kibri:t-a:t	= sulphate
kibri:t-i:d or *kibri:t-o:r*	= sulphide / Fr. sulfur
kibri:t-i:k	= sulphuric
kibri:t-o:z	= sulphurous / Fr. sulfureux

An explanation of this practice is that suffixes in scientific and particularly chemical terminology predominantly have specific meanings. Two or more suffixes may share a great deal of phonetic similarity; still this does not entail their designatory identity. Compare, for instance "phosphite" and "phosphate" which, for the reason just given, are given different renditions in Arabic, *fusfayt* and *fusfa:t* respectively (see, for instance, al-Khaṭīb, 1978, p.441). Chemical substances, compounds, elements, etc. have become so numerous, so varied, that in order to distinguish them it has become particularly necessary, when taken over by another language, to copy the words that denote them as faithfully as possible, so that scientific precision can be attained. This is particularly the case when a number of words have the same stem in common and differ only in the form of the suffix. In addition to the above examples, the point may be further illustrated by the combining form "chlor-/chloro-" which gives: chloral, chloramine, chlorate, chloride, chlorine, chloroform, chloromycetin, chlorophyll, chloropicrin, chloroplast, etc.

The same is certainly not equally true of ordinary vocabulary, where affixes are typically less specific in meaning than stems. Take for instance the suffix "-ful" in the following words:

peaceful	"characterized by" peace
pitiful	"deserving" pity
masterful	"having the qualities of" a master
meaningful	"full of" meaning
cheerful	"suggesting" happiness

It needs to be emphasized that the adoption of foreign elements in the type of N-constructions cited above is indicative of the fact that these elements cannot easily or conveniently be rendered through native means. As far as Arabic is concerned, the use of native equivalents for the remaining parts of the words containing these elements may thus be said to represent the most that can be done by way of avoiding the adoption of these words in their non-native character. As a matter of fact, words of the present type are more often than not fully adopted, i.e. borrowed as wholes, not only by Arabic, but by many other languages too; hence the label "International Scientific Vocabulary" (ISV) (see, for instance, Webster's New Collegiate Dictionary).

Attempts to create Hybrid N-constructions have also been made, though sporadically, outside the domain of chemistry. Consider the adoption of the suffix -logie, Eng. -logy, in the following words which occur in the Arabic translation of a French book on philosophy (the translator is ᶜAzīz al-Ḥabbābiyy, quoted by al-Shihābiyy, 1965, p.362):

fikru:-lo:jya:	vis-à-vis	Fr. idéologie
?ustu:r-lo:jya:		Fr. mythologie
qi:mu:-lo:jya:		Fr. axiologie
ka:?in-lo:jya:		Fr. ontologie

Unlike the case in chemical terminology, there seems to be no real justification for the use of the foreign suffix in these cases. To start with, the use of native elements in these cases does not present any difficulty, nor does it lead to any misrepresentation of the concepts involved. The Arabic word *ᶜilm* "science" is fully capable of standing for the suffix employed above. We could thus say *ᶜilmu l-?afka:r* "ideologie", *ᶜilmu l-?asa:ṭi:r* "mythologie", etc. It is obvious that the coiner of these constructions is trying to apply to Arabic what represents the case in French, which is not necessarily possible, particularly in the present situation. While formations of the type foreign word + native suffix may be quite frequent in Arabic as well as many other languages, cases where the order is reversed (i.e. native word + foreign suffix) are far too difficult to find. The former process is merely a matter of adopting a new lexical item — this explains why lexical borrowing is so common and automatic among languages. The latter, on the other hand, involves a great deal of complexity; it requires the familiarity of the native speakers with a new (foreign) structural pattern which may be entirely uncharacteristic of their own system. A pointer in this direction is Marchand's observation, in the study of English word formation, that native English affixes were added to borrowed (French) words almost immediately after they had been introduced, whereas foreign affixes did not appear with native words until much later and were far less numerous (Marchand, 1960, p.159). Besides, formations like *fikru:-lo:jya:* etc. are often rejected due to the mere fact that they are neither Arabic nor foreign. More often than not, foreign elements are met with antipathy arising from various cultural and linguistic considerations; there is always an element of resistance against them in the speakers' attitude. The fact that they are relatively new or unfamiliar is often sufficient reason for them to be opposed and stubbornly resisted even though sometimes they are apparently unexceptionable. Valerie Adams illustrates this fact with respect to English: throughout its history the verb "despise" has been nominalized in a number of ways, none of which has been regarded as fully or happily acceptable. There are thus: despite, despise, despisement, despiciency, despisal, despisery and despision (Adams, 1973, p.200).

When subjected to full arabicization (i.e. borrowed as wholes), words such as ideologie, mythologie, etc. are far more likely to gain currency than when partially arabicized. Indeed, the two words just quoted are frequently adopted by Arab writers, the Arabic versions being *ʔi:dyo:lo:jiyyah* and *mi:ṯo:lo:jya:*, respectively. In this case the borrowed words are not treated or analysed in terms of their constituent parts; rather they are taken as single wholes standing for certain concepts (see 4.4.3.1).

Some writers with a predilection for using foreign suffixes attempt to support their argument with what they mistakenly think to be a precedent established in the early stages of the language. For instance, a Tunisian scholar, Fūtah, proposes the use of the suffix *-u:n* (as against Latin -um or -ium) with native stems, e.g. *šams* "sun":*šamsu:n* "helium"; *qily* "alkali":*qilyu:n* "sodium". By so doing, the writer claims that he is following the example of such medieval physicans as al-Rāziyy and Ibn Sīna who, he argues, used the same suffix in words like *ʔafyu:n* "opium", *ʔasa:ru:n* "asarum", *ʔanisu:n* "anisum", *ġa:ritu:n* "agaricum" etc. (as reported by al-Suwaysiyy, 1976, p.13). However he seems not to have noticed that all these instances are fully arabicized forms of their respective Latin counterparts, i.e. the stems preceding the suffixes are not native.

3.4 Al-Murakkabu l-Mazjiyy (Mixed Compound)

Early Arab grammarians used the term *al-murakkabu l-mazjiyy* to refer to the combination into one lexical unit of words otherwise used independently. For example *baᶜlabakk*, the name of a town in Lebanon, from *baᶜl*, the name of an idol + *bakk*, the name of a person who worshipped that idol (see Ḥasan, 1963, vol.i, pp.270-1). However, the phenomenon is very limited and largely restricted to the names of persons and places. Moreover, most of the examples cited by the grammarians are in fact constructions of foreign origin which are reatomized into their original constituents, e.g. *Si:bawayh*, the name of a very celebrated Arab grammarian (d.180 AH), is originally Per. si:bu:ya "the perfume of an apple" from si:b "an apple" + bu:ya "perfume" (cf. Steingass, 1892, p.714). Sometimes the compound construction represents a combination of a native constituent and a foreign one, the ultimate unit being coined by the foreign language rather than Arabic. For instance, *sila:ḥda:r* "arms-bearer" from Ar. *sila:ḥ-* "weapon" + Per. -da:r "keeper, possessor" (from da:štan "to possess").

There are also instances of relative (nisbah) adjectives which are compounded out of two (full) nouns, the first being in the status constructus governing the second in the genitive, thus forming an annexative structure. For example, *da:raquṭniyy* belonging to *da:ru l-quṭni* the name of a quarter in Baghdad, *dayraᶜa:qu:liyy* a native of *dayru lᶜa:qu:li* the name of an old city in the southern part of Baghdad, *ḏunnu:niyy* a descendant of *ḏu: n-nu:n* the name of a family in Spain, etc. (see Wright, 1967, vol.1, pp.21-2). However there are cases where the two constituents underlying the murakkab mizjiyy are both given the nisbah suffix, i.e. *-iyy*. Thus a person belonging to the city named *ra:m hurmuz* is described as *ra:miyy hurmuziyy*, as opposed to the compound form *ra:mahurmuziyy* which is also recorded in the early literature (see Ḥasan, 1963, vol.4, p.557; and p.67 above).

Modern coinages based on the principle illustrated above are also detectable. A case in point is the word *barma:?iyy* "amphibious"(33) which is coined from *barr*(34) "land (as opposed to sea)" and *ma:?* "water". The fact that the preceding coinage has gained rapid acceptance and is considered as a better alternative than the corresponding double-nisbah formation *barriyy ma:?iyy* may be accounted for by the fact that besides being more economical than the latter, it is also equally understandable as the underlying constituents are still distinguishable. There is also the fact that being a single morphological unit, it can easily be handled in grammatical and syntactic processes. It would be more convenient for instance to render the phrase "amphibious animals" as *al-ḥayawa:na:tu l-barma:?iyyah* or simply *al-barma:?iyya:t* than as *al-ḥayawa:na:tu l-barriyyatu l-ma:?iyyah*; notice also the use of the definite article in both cases.

One of the problems arising from the combining of two words together is the possible occurrence of more-than-two-consonant clusters, which has already been said to be inadmissible in Arabic. It is for this reason that objection is often expressed against formations such as *xalfmiḥwariyy*(Muḥammad, 1965, p.263) "post-axial" (*xalf* "post-" + *miḥwariyy* "axial") where a three-consonant cluster, -lfm-, is created. By way of getting round this problem, some writers have resorted to the use of a junctural vowel in order to avoid such clusters. Thus the word just quoted would (according to al-Baᶜlabakkiyy, 1973, p.711) be rendered *xalfi:mihwariyy*. It may have been noticed that, etymologically speaking, the inserted vowel is in fact derived from the nisbah-suffix *-iyy* as would appear at the end of the first word in the double nisbah-formation (or co-ordinated attributes) *xalfiyyun miḥwariyy*. Here are some more examples of similarly featured words as presented in al-Baᶜlabakkiyy's dictionary:

šafawi:sinniyy	labiodental	(p.508)
?anfi:bulᶜu:miyy	nasopharyngeal	(p.604)
baᶜdi:taxarrujiyy	postgraduate	(p.711)
?ama:mi:jabhiyy	prefrontal	(p.718)
nawawi:ḥara:riyy	thermonuclear	(p.963)

Formations of the same nature were earlier proposed by Maḍhhar, from whose al-Nahḍah Dictionary we quote the following sample:

Murakkab mazjiyy	Set combination	
al-quwwatu ṭ-ṭardi:markaziyyah = centrifugal force	*al-quwwatu ṭ-ṭa:ridatu ᶜani l-markaz*	(vol.1, p.192)
al-?iltiha:bu l-kabidi:ṭiha:liyy = hepato-splenitis	*?iltiha:bu l-kabidi wa ṭ-ṭiha:l*	(vol.1, p.615)
al-ba:ḥa:tu l-ḥawli:wiᶜa:?iyyah = perivascular spaces	*al-ba:ḥa:tu ḥawla l-wiᶜa:?iyyah*	(vol.1, p.1570)
at-tarki:bu r-ruba:+i:qadamiyy = tetrapodal structure	*at-tarki:bu r-ruba:ᶜiyyu l-?aqda:m*	(vol.2, p.2231)

We may also include as instances of murakkab mazjiyy cases where the negative particle *la:* is compounded with another constituent, both being fully represented.(35) In such cases Ar. *la:-* does a similar function to that of such English negative prefixes as: a-, an-, anti-, in-, un-, non-, etc. and the suffix -less. We thus find:

la:jana:ḥiyy	apteral
la:tana:ẓur	asymmetry
la:ma:?iyy	Fr. anhydrique
la:sa:miyyah	anti-semitism
la:fana:?iyyah	indestructibility
la:?ana:niyyah	unselfishness
la:qišriyy	skinless

Notice that formations such as these may also be used with the definite article, e.g. *al-la:jana:ḥiyy*. It was mentioned above that instances representing this development are also detectable in medieval Arabic. In the present stage, however, the phenomenon is far more widely applied. Evidence for this statement may be derived from the fact that the applicability of *la:-* has now become such that other elements in the language which may occur more or less interchangeably with it, but which may be used only separately, are gradually being ousted by it. Such elements are represented by *ᶜadam* "non-, dis-, etc." and *ġayr* "other than, different from, etc." These elements are normally used with a following noun in the genitive to form an antithetical construction, e.g. *ġayru šarᶜiyyin* "unlawful". However, there is a growing tendency among the speakers, especially in technical contexts, to replace *ġayr* or *ᶜadam* by *la:-* and say, for instance: *la:šarᶜiyy*, *la:qa:nu:niyy* "illegal", *la:ᶜunf* "nonviolence" etc. A term like "decentralized authority" is almost invariably *sulṭatun la:markaziyyah* (as opposed to *sulṭatun ġayru markaziyyah*). Likewise, *ᶜadam* has largely given way to *la:-* in *al-la:muba:la:h* "nonchalance", *al-la:mas?u:liyyah* "irresponsibility", *al-la:šuᶜu:r* "the unconscious", *al-la:di:niyyah* "irreligion", etc. It could be stated in general that the more specialized or technical the antithetical term, the higher the priority of *la:-* over *ᶜadam* and *ġayr*. For instance, when the aim is to express an ideology or principle (*mabda?*), the term "non-violence" is almost always rendered as *al-la:ᶜunf*, e.g. *?an: ?u?minu bi-mabda?i l-la:ᶜunf* "I believe in the principle of nonviolence"; otherwise, *ᶜadamu l-ᶜunf* is equally possible, e.g. *?ana: ?ufaḍḍilu ᶜadama stiᶜma:li l-ᶜunf* "I prefer not using violence".

Double nisbah-formations

We have pointed out that a formation like *xalfi:miḥwariyy* is underlain by two co-ordinated adjectives, both terminated by the nisbah-suffix *-iyy*, viz. *xalfiyyun miḥwariyy*. It may be useful to bring into focus the point that, as far as actual practice is concerned, the latter construction represents what is almost always the case in the language. Formations represented by the former are still in the form of proposals and have no existence beyond the dictionaries or articles where they are put forward as suggestions, reflecting their authors' attempt at developing more adequate techniques of lexical creation. Thus, foreign compound terms are commonly rendered through the use of co-ordinated nisbah-adjectives. Here are some examples:

electrochemical corrosion	*haṯṯun ki:mya:?iyyun kahraba:?iyy*
mechanical electroplating	*tila:?un kahraba:?iyyun mi:ka:ni:kiyy*
electrocapillarity	*al-xa:ṣṣatu š-šaᶜriyyatu l-kahrabiyyah*
cholecystogastrostomy	*tafa:ġumun ṣafra:wiyyun maᶜidiyy*
cathodophosphorescence	*al-fusfo:riyyatu l-ka:ṯo:diyyah*
artificial earth satellite	*at-ta:biᶜu l-?arḍiyyu ṣ-ṣina:ᶜiyy*
radiotelephony	*at-talifo:niyyatu l-la:silkiyyah*

Notes to Chapter 3

1 As quoted by Jirjis (1961, p.63): *"?inna l-ᶜaraba talja?u lin-naḥti ?iḏa: kaṯura s-stiᶜma:luhum lil-kalimatayni ḍammu: baᶜḍa ḥuru:fi ?iḥda:huma: ?ila: baᶜḍi ḥuru:fi l-?uxra:"*

2 Ibn Fāris (1964, p.271): *"al-ᶜarabu tanḥatu min kalimatayni kalimatan wa:ḥidatan wa huwa jinsun mina l-?ixtiṣa:r"*

3 Another term "crossing" is also used to describe "the play of two elements acting on each other so as to bring about the formation of a third element which partakes of both"; see Pei, 1966, p.58.

4 The term "contamination" has already been used with reference to Arabic by Henri Fleisch (1956, p.124).

5 A fuller account of this word and its underlying forms is given on page 63.

6 This is in fact a metaphorical sense of *jadub-* "to be or become arid (of soil, land)".

7 See Steingass, 1892, p.1081. It should be noted that Steingass symbolizes Persian long vowels differently, e.g. /a:/, as in the word just quoted, would be /ā/. We shall, however, continue to apply our own system when quoting him again.

8 Ibn Fāris, 1963, p.271; see also al-Suyūṭiyy, n.d., vol.1, p.482.

9 Ibn Fāris, 1946, vol.1, p.331.

10 Ibid., vol.3, p.351.

11 al-Suyūṭiyy, n.d., vol.1, p.483.

12 Ibid., p.484.

13 Ibid., p.484.

14 Ibid., pp.484-5.

15 A modern writer, Ramsīs Jirjis has pointed out what may best be described as generalizations based on the recurrence of certain features. However, the inconsistencies and exceptions are too many to belittle. Another scholar, al-Shihābiyy, sees as a general rule the combining into one unit (mostly of the measure Fa8LaL(ah)) of the first two radicals of both components; but again this rule is also often violated, as the writer himself admits. See his article in MM^cI^cA, 1959, pp.545-7. Cf. also Ghazāl, 1978, p.22; also Maḍhhar, "Tajdīd", pp.48-9.

16 This is true not only with regard to nouns and verbs, but also in the case of prepositions and particles, apart from a few exceptions in the last category. These exceptions are represented by the article *al-*, the prepositions *bi-* "with, etc.", *li-* "for, etc.", the future-denoting particle *sa-* "will", and the conjunction *fa-* "then, and so", which are formally marked by being tacked on to the following word. See Zgusta, 1971, p.160; Beeston, 1970, p.30.

17 MML^cA, vii, 1953, p.201: *"an-naḥtu ḍarbun mina l-?ixtiṣa:r wa huwa ?axḏu kalimatin min kalimatayni fa-?akṯar"*.

18 The committee consisted of the academy members: Ibrāhīm Ḥamrūsh, Maḥmūd Shaltūt, Aḥmad Zeki, Muṣṭafa Naḍhīf and ^cAbd al-Qādir al-Maghribiyy; see MML^cA, vii, 1953, p.201. See p.203: *"wa naḥnu naqu:lu bi-jawa:zi n-naḥti fi: l-^culu:mi wa l-funu:ni lil-ḥa:jati l-mulihhati ?ila: t-ta^cbi:ri ^canha: bi-?alfa:ḏin ^carabiyyatin mu:jazah"*; and p.158: *"yuja:zu n-naḥtu ^cindama: tulji?u ?ilayhi d-ḍaru:ratu l-^cilmiyyah"*.

19 Notice that the second underlying constituent in this case is an arabicized form.

20 MML^cA, vii, 1953, p.204.

21 By way of illustrating the adoption of naḥt as a form of lexical adaptation, some writers have resorted to evidence from Colloquial Arabic; see, for instance, Zaydān, "al-Falsafah", pp.71-87.

22 al-Ḥuṣriyy, 1958, p.146.

23 al-Ba^clabakkiyy, 1973, p.877; cf. also al-Ḥuṣriyy's version of the same coinage, p.71 above.

24 Maḍhhar, "al-Nahḍah", 2, p.1247; and 1, p.686.

25 Cairene Academy, MML^{c}A, vii, 1953, p.204.

26 al-Ba^{c}labakkiyy, p.258.

27 al-Kawākibiyy, p.40.

28 al-Ḥuṣriyy, p.145.

29 Maḍhhar, al-Nahḍah, vol.i, p.460.

30 Ibid., vol.1, p.461.

31 Ibid., vol.2, p.1483.

32 It should be mentioned however that the case we are referring to here is an instance of Full N-construction.

33 This coinage is due to Anīs al-Khūriyy, as mentioned by al-Husriyy, 1958, p.144.

34 The second */r/* of this word is elided in the compound construction due to the fact that it is not permissible in Arabic to have a three-consonant cluster.

35 Cf. the partial representation of *la:* in N-constructions, p.67 above.

4 Arabicization : a method of lexical expansion in Arabic

Introduction

Among the methods of lexical expansion adopted by Standard Arabic is the one traditionally known by the name of *at-Taᶜri:b* (Arabicization), whereby linguistic elements of non-native origin are taken over and used in the language. This phenomenon, which can be traced as far back in the history of Arabic as pre-Islamic times, has always been a matter of dispute among Arab philologists, grammarians and writers. This is especially the case with regard to such issues as: the need for borrowings and in what sphere of the language; the status of the borrowed word in Arabic; the significance of arabicization among other methods of lexical expansion; and the impact of this phenomenon on the language as a whole.

The preceding questions constitute the focus of our concern in this chapter, with special emphasis on investigating the practicability of this method and the role it plays in the language of science and technology. Our attention will be mainly directed towards examining the receptivity of Arabic to non-native elements, the way it handles them, the various attitudes held towards them and their role in the language throughout its history. We hope to find out to what extent Arabic has resorted to foreign borrowings, under what circumstances and in what domains – and to look at various views and theories put forward and the sort of impact they have had on the state of affairs.

As has been done previously, our investigation of these issues will be illustrated and supported with what may be considered as observations of universal applicability, i.e. phenomena that are equally applicable to other languages.

4.1 Preliminaries and definitions

Before we start, we need to acquaint ourselves with a number of notions and terms that keep cropping up in any discussion on lexical expansion in general and linguistic borrowing in particular. There are also a number of definitions and terms commonly employed by Arab writers which will be met frequently in the course of our discussion and which therefore call for some consideration.

The process whereby a particular language incorporates in its vocabulary words from another or other languages is technically designated by such terms as "borrowing", "adoption", "loaning", etc. Greenberg (1957, p.69), for instance, defines borrowing as "the acceptance in one language of a form, in both its sound and its meaning aspects, from another language; though usually with both phonetic and semantic modifications". However, "borrowing" as well as the two other terms mentioned above is sometimes criticized as unsuitable. The main criticism arises from the incompatibility of their logical (or literal) connotations and their conventional (or technical) meanings. In his analysis of borrowed vocabulary in English, Matthews, for instance, argues that: (1979, p.47)

> when one borrows something it is on the strict understanding that it will be returned, and whoever thought of returning a word to the French or Italians or anyone else from whom we had filched it? For one thing we haven't deprived them of it, and for another they might not want it after what we have done to it, for we are apt to treat our borrowings rather roughly.

However, since in linguistic discussions on borrowing these terms are not taken in their literal meaning and have been established by use rather than by logic, we think it useful, as Matthews himself does, to retain them in our study. Words or items involved in the process under discussion may thus be referred to as borrowings, loans, loanwords, loanforms or adoptions.

As far as Arabic is concerned, the term *Taᶜri:b* is used by Arab writers in two different, though related manners. On the one hand, it is used to designate the process whereby foreign words are incorporated into the language, as mentioned above. On the other, it is also applied in a broader sense to refer to all types of methods involved in the rendering of foreign terminology into Arabic, whether by arabicization, in the sense just explained, or by a variety of other methods including translation, semantic extension of already existing native elements, revival of old vocabulary, etc. In other words, the process in the latter sense represents the various efforts aimed at modernizing Arabic and enriching it with new, especially scientific and technical, vocabulary; hence the name of the recently established organization "The Permanent Bureau of Co-ordination of Arabization" (PBA) in Rabat. From our point of view, however, the term "arabicization" will be used in its former sense only. It is also frequently the case in our discussion that this term is qualified by another one, "analogical arabicization" (*at-Taᶜri:bu l-Qiya:siyy)*, in which case it is specifically used to refer to the process whereby loanforms are modified in such a way as to be in full harmony with the structural character of Arabic words; that is, that they assume native patterns and may be used for further derivation like any native elements. Otherwise, if borrowings are not so modified, i.e. if they retain their foreign character, the process is designated by the term "arabicization" alone.

The Arabic literature on borrowing also contains a number of distinctions that are relevant to our discussion. According to one of these distinctions, a word used by the Arabs may be broadly categorized on the basis of its origin, either as *ʔaṣi:lah* "of Arabic origin (*ʔaṣl*)" or *daxi:lah* "of foreign origin". The same distinction is sometimes indicated by the terms *ᶜarabiyyah* "Arabic" and *ʔaᶜjamiyyah* "non-Arabic", respectively. (The term *ʔaᶜjamiyy* was originally used to designate a non-Arab, especially a Persian, and was then applied to foreign words of Persian origin. Nowadays it is sometimes used to refer to any borrowing of any origin.)

Non-native words are further subcategorized into *muᶜarraba:t* and *muwallada:t*. The former represent those borrowings which were taken over and used during the classical period, i.e. pre- and early Islamic times. The latter are those that have entered the language subsequently, in the post-classical period. Other considerations underlying this distinction are discussed in 4.3.2; meanwhile it may be useful to point out that the above use of the term *muwallada:t* "produced, generated" (w-l-d "to give birth to") is paralleled by an analogous application of the same term in social life. Early Arabs referred to those members of their society who were not of Arabic descent as *al-muwalladu:n*. Thus a *muwallad* person was one who was born and raised among the Arabs but one or both of whose parents were not pure Arabs. Similarly a *kalimah muwalladah* is a non-Arabic word which has found its way into the language as a result of the social, cultural or other sorts of contact between the Arabs and other nations. al-Aṣmaᶜiyy is quoted by al-Suyūṭiyy ("al-Muzhir", vol.1, p.304) as saying that "[the word] *an-niḥri:r* [intelligent man] does not belong to the Arabic language, it is a *muwalladah*". Words so designated, though they may be in current use, are generally considered as unqualified to be included as part of the Arabic vocabulary, mainly due to the fact that unlike the *muᶜarraba:t* they were not used by the classical speakers, the "eloquent of the Arabs" (*fuṣaḥa:ʔu l-ᶜarab*) (see Wāfi, p.199).

Periods of citation

When early Arab philologists and language codifiers set out to collect their material, their work was characterized by a remarkable degree of caution and meticulousness. They were at pains to ensure that what they recorded was truly representative of the pure Arabic character — so much so that they were extremely careful in choosing their informants, who were mostly Bedouin Arabs (*ᶜarabu l-ba:diyah*), the inhabitants of the desert, rather than town dwellers (*ᶜarabu l-ʔamṣa:r*). The former were, and indeed still are, felt to be superior to the latter with regard to their intuitive knowledge and mastery of Classical Arabic in its purest form. This is mainly due to the fact that, unlike urban residents, those Bedouins had no or very little contact with non-Arabs, which accounts for the fact that their dialect remained unexposed to any substantial foreign influence. Moreover, some tribes were considered more reliable in this respect than others. We are thus told that while the tribes of Quraysh, Qays, Tamīm, Asad, and Hudhayl were taken to be the most reliable sources, those of Lakhm, Judhām, Quḍāᶜah, Ghassān, etc. did not enjoy the reliability because they were bordered by non-Arabic speaking peoples, e.g. Persians, Copts, Greeks, etc. (see al-Jawhariyy d.1002).

The caution taken by early codifiers had also a temporal dimension. For not only did they confine their investigation to particular areas or communities but also to particular periods, in which, it is argued, the Arab tongue remained flawless, unimpaired and immune from the confusion which later periods came to witness due to the influence of foreign languages and the spread of "faulty speech" (*al-laḥn*). Thus, by the end of the second Islamic century, they ceased to quote from the Arabs living in capital cities, whereas they continued to rely on the Arabs of the desert up to the middle of the fourth century. The terms *ᶜuṣu:ru l-ʔistišha:d* "periods of citation" and *ᶜuṣu:ru l-ʔiḥtija:j* "periods of attestation" are commonly used by modern writers when referring to the type of Arabic characteristic of these periods.

It should be mentioned that in addition to the language of the desert, the Qurʔān, the paragon of purity and eloquence, represents one of the most important models early philologists adopted in their attempt to codify the language. There are also the sayings and speeches of the Prophet as well as pre- and early Islamic poetry and proverbs. The desert, however, was regarded as a linguistic laboratory whose inhabitants – the standard bearers of purism and eloquence – were often called upon to arbitrate linguistic disputations (see Chejne, p.40).

4.2 The phenomenon of arabicization: its history and development

4.2.1 Arabicization in the pre-Islamic period

It is an established fact that when two or more languages are brought into contact with one another they tend to undergo a process of inter-influence, especially at the lexical level. The type and degree of contact may vary from one case to another, but the fact always remains that all languages are susceptible to external linguistic influence, though the extent of susceptibility may be said to be more or less determined by certain factors, e.g. whether the languages involved are genetically related or not.

Arabic in this respect is by no means different from other languages. Prior to the advent of Islam, as well as after it, there have been various types of contact between Arabs and other nations. History tells us that the people of Arabia, especially the western region of it known by the name of al-Ḥijāz, were active traders, who, as a natural result of their frequent journeys outside their homeland, developed strong commercial, economic and cultural, as well as family, ties with the people with whom they came into contact (see Zaydān, "al-Lughah", pp.28-32). Present-day Arabic is generally said to trace its origin mainly to the language spoken in this area.

Another factor leading to the contact betwen Arabic and other languages is the fact that the Holy Kaᶜbah in Mecca was, as it still is, the destination of pilgrims from different parts of the world, Indians, Persians, Abyssinians, etc. Among the traces those pilgrims left behind were words belonging to their own languages, some of which later found their way into the most literary and most representative sectors of Arabic, namely the Qur?ān and classical poetry, as we shall see below.

Geographical proximity between Arabic and other Semitic languages, Ethiopic, Hebrew and Aramaic, which facilitated other sorts of contact, was another factor accounting for the infiltration of borrowings from these languages into Arabic and vice versa. Linguistic research has revealed for instance that Aramaic was the source from which Arabic derived many of its early philosophical and metaphysical vocabulary as well as words pertaining to industry and other aspects of urban life, e.g.*šayṭa:n* "Satan, devil", *sikki:n* "knife", *sa:riyah* "column of stone or baked bricks", etc. (see Wāfi, pp.127-8).

Contact between Arabic and a number of non-Semitic languages, mainly Persian (Pahlavi) and Greek, can also be traced to pre-Islamic times. The inter-influence between the former and Arabic has been such that a great majority of the vocabulary of Modern Persian can easily be proved to be of Arabic origin, and on the other hand non-native elements of Persian origin are not difficult to find in Arabic. Instances of such elements were used by such pre-Islamic poets as Ṭarafah ibn al-ᶜAbd, Imru? al-Qays, ᶜAdiyy bin Zayd, al- Aᶜsha, etc. (see al-Jawāliqiyy, d.540 AH, pp.102, 302, 71, 128 respectively).

Evidence of more or less direct contact between Arabic and Greek, especially in the period immediately preceding Islam, is also available. "At that time Byzantine influence was supreme in Syria and Palestine, and the Arab confederacy of Ghassān, which acted as a buffer state between the Byzantine Empire and the desert tribes ... was a channel whereby Byzantine influence touched the Arabs at many points." (Jeffery, p.14). According to Nicholson, Greek culture had established itself in Egypt, Syria and Western Asia since the time of Alexander's conquest, i.e. since 322 BC (Nicholson, p.358; Baqir, p.589).

In Mesopotamia, the city of Ḥarrān was another centre of Hellenism. Major contributions to the diffusion of Greek culture in the area came from the inhabitants of that city, Syrian heathens who were capable of speaking Arabic. They were almost the only translators of the writings of such ancient masters as Aristotle, Galen, Ptolemy, etc. (Nicholson, p.358). Their (Syriac) versions were, however, afterwards retranslated into Arabic, which accounts for the existence in the latter of certain features that are not inherent in it but are rather borrowed from Syriac.

The Byzantine court in Constantinople (al-Qusṭanṭiniyyah) is also reported to have been frequented by a number of Arab poets, among whom was the aforementioned Imru? al-Qays, the author of the oldest and most famous of the Muᶜallaqāt (the name given to seven old Arabic poems which were suspended in the Kaᶜbah in Mecca). As would be expected, foreign borrowings of Byzantine origin are not uncommon in his poetry (see al-Jawāliqiyy, pp. 74, 201, 227, 319, 401).

4.2.2 Arabicization in the Islamic period: Qur?anic borrowings

The fact that the Qur?ān contains a number of originally non-Arabic elements was fully recognized and frankly admitted by the earliest circle of exegetes and companions of the Prophet, among whom was the Prophet's cousin Ibn ᶜAbbās (d.68 AH). He declared the non-nativeness of such words as:(1)

sijji:l	lumps of baked clay
al-miška:t	a niche in a wall
al-yamm	sea, flood
aṭ-ṭu:r	Mount Sinai
?istabraq	silk brocade

There were also other advocates of the same view, e.g. ᶜIkrimah, ᶜAṭā?, Saᶜīd bin Jubayr, etc. (see al-Suyūṭiyy, "al-Muzhir", 1, p.268). However, the existence and status of foreign words in the Qur?ān has been one of the most controversial questions among Muslim writers, especially those of early Islamic centuries. In contrast to the position taken by the above authorities, a large number of their successors have strongly denied the existence of any borrowings in the Holy Book. Their fundamental argument in this respect is that the Qur?ān in several passages refers to itself as an Arabic Qur?ān, and to support their argument they cite such verses as *?inna: jaᶜalna:hu qur?a:nan ᶜarabiyyan* (43:3) "verily we have made it an Arabic Qur?ān"; *wa kaḏa:lika ?anzalna:hu ḥukman ᶜarabiyyan* (13:37) "thus have we revealed it a decisive utterance in Arabic"; *bi-lisa:nin ᶜarabiyyin mubi:n* (26:195) "in plain Arabic speech", etc. Thus, arguing against the existence of non-Arabic words in the Qur?ān, Abū ᶜUbaydah (quoted by al-Jawāliqiyy, p.52) has this to say: "He who claims that there is in the Qur?ān anything other than the Arabic tongue has made a serious charge against God." More or less similar attitudes are held by al-Shāfiᶜiyy, Ibn Jarīr, al-Qāḍi Abū Bakr, Ibn Fāris, etc. (see al-Suyūṭiyy, 1967, vol.2, p.105).

Statements based on an objective and realistic approach to the problem have not been lacking. For instance Abū ᶜUbayd al-Qāsim bin Salām is quoted to say: "Certainly these [borrowed] words are originally not in the Arabs' speech ... [but] then the Arabs committed them to their tongues and arabicized them ... so from this point of view they are Arabic [but ultimately] of foreign origin" (quoted from al-Jawāliqiyy, p.53). This latter view seems to have gained acceptance among modern writers, and is therefore put forward almost in every discussion of the phenomenon.

We have avoided giving a detailed analysis of the conflicting views involved in this subject, nor do we intend to give an exhaustive list of the foreign vocabulary in the Qur?ān, since these issues have been dealt with by several writers;(2) however there are certain observations which we find of sufficient interest to discuss here as they have much bearing on the question of incorporating foreign vocabulary in general, as well as the validity of the attitudes adopted towards this phenomenon.

The conflicting views of early authorities concerning the existence of loanwords in the Qur?ān correlate fairly closely with the discrepancies exhibited by some modern writers dealing with the question of arabicizing foreign scientific and technical terminologies. The underlying factor in both situations seems to be the type of attitude held towards the purity and sufficiency of Arabic as a medium for the divine revelation in the former case and the dissemination of scientific knowledge in the latter. More often than not, some Arab philologists have expressed the view that Arabic is so self-sufficient that there is no need for it to borrow words from other languages, and that the latter fall short of attaining to its level of wealth and perfection. Being the language of the Qur?ān, the final and most perfect of divine revelations, it must then, it is maintained, be the most perfect of all languages. According to Ibn Fāris, "if there is anything in it [the Qur?ān] from a language other than Arabic, that would raise the suspicion that Arabic fell short of producing it, so that it had to come in languages they [the Arabs] did not know" (quoted by al-Suyūṭiyy, 1967, 2, p.105). Statements to this effect are in fact still sometimes being made by more recent wriers as we shall see below.

While acknowledging the richness of Arabic, one cannot help pointing out that statements like the one just quoted are guided not so much by objective understanding and a realistic approach to the problem in question as by religious zeal and unverified impressionistic judgement. For it has been proved beyond all doubt that the Qur?ān does contain a number of originally non-Arabic elements belonging to several languages, both Semitic and non-Semitic. In addition to the instances cited above, we thus find *ṣira:ṭ* "path, way" from Latin strāta (see Jeffery, pp.195-6); *misk* "musk" from Sanskrit muska (Webster, p.759; Onions, p.598); *qami:ṣ* "shirt" from Greek καμισιον (Jeffery, p.243); etc. Among the cognate languages to which some of the Qurā?nic borrowings belong are Ethiopic *al-ḥabašiyyah*, Syriac *as-surya:niyyah*, Hebrew *al-ᶜibriyyah*, and Nabataean *an-nabaṭiyyah*. Al-Suyūṭiyy's list also includes borrowings from Hamitic languages: Berber *al-barbariyyah*, Coptic *al-qubṭiyyah* and Negro *az-zanjiyyah* (1438 AH, p.3).

What some investigators seem to have failed to realize is that the occurrence of foreign borrowings in the Qur?ān or any other sector of language need not imply any unfavourable value judgement about Arabic. The existence of words from linguistic origins other than that of a particular community or nation is a natural outcome of the various types of contact and interaction among the co-habitant nations of the world (see also 4.3.2).

It is worth our while to mention here that the accounts given by some early scholars of the Qur?ānic non-native vocabulary reveal a certain degree of inaccuracy and a lack of adequate knowledge of the languages constituting the sources of Qur?ānic borrowings. In a number of cases some Qur?ānic words have been designated as foreign borrowings, whereas in actual fact they are purely Arabic but are thought otherwise due to the fact that they are introduced in rare contexts or their usual meaning is not clear on the surface. According to al-Suyūṭiyy's authorities for instance the following words derive from non-Arabic origin:

hayta laka	mistakenly thought to be Coptic (Qur?ān, 12:23)
taḥta	mistakenly thought to be Nabataean (19:24)
?iblaᶜi:	mistakenly thought to be Indian or Ethiopic (11:46)
?axlada	mistakenly thought to be Hebrew (7:176)
ḥaṣab	mistakenly thought to be Negro (21:98)

However, a careful investigation of these words would leave no doubt as to their native origin (for further examples and a detailed account, see Jeffery, pp.32ff).

It remains to be added that the occurrence of foreign words in the Qur?ān provides us with clear evidence of the fact that those words had existed in pre-Islamic times, prior to their appearance in the Qur?ān, which leads us to a further conclusion, namely that lexical borrowing is not a novel or unfamiliar phenomenon in present-day Arabic.

4.2.3 Borrowing as a consequence of cultural interaction between Muslim Arabs and other nations

The process of cultural interaction between Arabs and other nations became more active than ever before with the advent of Islam. The radical changes introduced by the new religion in the ideological, social and conceptual make-up of the previously heathen society were now powerful forces which helped the newly born nation orientate itself with regard to other nations and the world around it. Regarding themselves as the carriers of a new divine message, Muslims found it their duty to deliver that message to other peoples, no matter how far away or who they were. Thus, through conquest, immigration, intercommunication, etc. they brought themselves into contact with other nations far more intimately than they had done in pre-Islamic times.

As a natural concomitant of these circumstances, various types of mutual influence took place between Muslims and the indigenous inhabitants of the countries they reached. Of course, language was one of the susceptible and important domains for such a process. It may be argued that in such circumstances, Arabic, being the carrier of the new message and the language of the conquerors, was itself influencing other languages not influenced by them. It cannot be ruled out, however, that it was not entirely immune to the various channels of external linguistic influence, particularly loanwords. A substantial argument in favour of the conclusion is the fact that the language of any nation is closely interrelated with its culture, and that when an item relating to one culture is borrowed by another, it is often accompanied by its name or label (see also 4.3.2). It cannot be denied that cultural contact with other nations had a far-reaching effect on the development and enrichment of the scientific thought of early Muslim Arabs. The latter, in turn, did a great service to the world by accumulating the early scientific works of the Greeks, Indians, Persians, Chaldeans, etc., which together with their own abundant contributions and after having been meticulously studied, expounded and elucidated, came to be valuable sources of knowledge to scholars and students of science from various parts of the world (see Ṭāha, 1976, pp.339-89). It may thus be said that the linguistic inter-influence between Arabic and other languages was more or less a reflection of the cultural interaction between Arabs and non-Arabs.

An important aspect of the cultural contact between Muslim Arabs and other nations is represented by the successful and fruitful efforts of the former at translating and studying foreign sciences. Historians agree that this movement started in the first Islamic century under the Umayyad Prince Khālid bin Yazīd, who procured the translation of Greek and Coptic works on chemistry and who himself wrote three treatises on the same subject. The movement was also greatly encouraged by the Caliph Marwān bin al-Ḥakam and his son ᶜAbd al-Malik bin Marwān, as well as the latter's son, Hušām.

However, it was in the Abbasid period that such studies received the greatest impetus. During the reign of Caliph al-Manṣūr (745-75), works on logic and medicine were translated from Pahlavi by Ibn al-Muqaffaᶜ and others. Caliph al-Rashīd, too, showed great enthusiasm for promoting scientific research; and so did his son al-Ma?mūn (813-33), who was interested in Greek learning and established his famous House of Wisdom (Bayt al-Ḥikmah) in Baghdad, which included a bureau of translation. (3) It was due to his great encouragement and sponsorship that many works on astronomy, mathematics, geography, philosophy and medicine were rendered into Arabic (see Chejne, p.69). His ample rewards attracted a large number of translators to whom he set the task of translating books on various scientific fields (Nicholson, pp.358-9).

It is not our concern here to go into further details about the movement of translation in Islam and its impact on Arabic civilization and the world in general. (4) The remarks made above have been intended to lead us up to a point of direct pertinence to our subject proper, i.e. lexical borrowing, on which we shall focus our attention in the following discussion.

One of the outcomes of the movement discussed above was the flow into Arabic of a large number of foreign terms to convey the non-native concepts and the ideas they stood for. The process seems to have been practised unreservedly in its initial stages. It is a widely observed phenomenon in early translations that foreign words, mostly Greek, are simply transliterated into Arabic letters. This may be accounted for by the fact already mentioned that the first translations were not done by native speakers of Arabic, rather they were the works of Syriac-speaking translators who, in spite of their good command of Greek, were not equally efficient in Arabic (see Browne, pp.28-9). Many of these borrowings were later either replaced by native equivalents or modified in accordance with the characteristic structure of the language. In some cases other foreign but fully naturalized substitutes were used. The following are some cases in point:

dira:xma:	drachma	*dirham*
?astro:no:mya:	astronomy	*(ᶜilm) al-falak*
jiyo:maṭri:qa:	geometry	*(ᶜilm) al-handasah* (5)
ri:ṭo:ri:qa:	rhetoric	*(fann) al-xita:bah*
manja:ni:qo:n	mangonel	*manjani:q* (6)
so:fso:ṭi:qa:	sophistry	*al-muġa:laṭah* or *as-safsaṭah*
?ariṯma:ṭi:qa:	arithmetic	*(ᶜilm) al-ḥisa:b*
?abi:di:mya:	epidemic	*al-wa:fidah*
di:na:ryo:s	denarius	*di:na:r*

It could thus be said that the process of borrowing was first taken as a short-cut, especially by unqualified translators who must have found it a handy procedure whenever an appropriate Arabic equivalent was not readily available. The subsequent replacements or modifications made by Arabic-speaking translators and writers represent their attempts at bringing the new terminology into line with the native structure as far as possible. It is noticed that foreign borrowings that have been fully naturalized, i.e. analogically arabicized, e.g.*handasah* and *manjani:q*, have been happily incorporated into the language and are treated on a par with native elements. Non-analogical borrowings, on the other hand, have proved far less capable of survival and are always threatened by elimination (see 4.4.1 ii).

The replacing of foreign borrowings by native equivalents or modifying them in accordance with the native system constitutes a twofold practice currently involved in the process of adaptation and modernization nowadays being witnessed by SA. However, this need not imply the nonexistence of borrowings that are not (or have not yet been) analogically arabicized, especially in the case of scientific vocabulary, as we shall see in due course.

In addition to the movement of translation there have been other channels through which loanwords found their way into Arabic. These are represented by other aspects of contact and interaction discussed earlier between Arabs and other peoples. Needless to say, in the present time these forms of contact are far more varied and more intimate due to the spread of knowledge and the development of communication among the various parts of the world.

4.3 Borrowing as viewed by Arab writers

No sooner had Arab philogists, grammarians and lexicographers been aware of the existence of foreign words in their language than they began to inquire into the origin, status and characteristics of these elements vis-à-vis the native vocabulary of the language. A brief review of the discussions that have revolved around the problem may help us form a number of conclusions with regard to such matters as:

(a) the theoretical principles on which attitudes towards loanwords have been based;
(b) the impact of these views on the question of lexical expansion as a present-day problem; and
(c) the practicability and extent of applicability of the principle of Analogical Arabicization.

There are also a number of other relevant issues to which we shall devote part of our discussion.

4.3.1 The phenomenon of borrowing in Arabic literature

The study of borrowing as a linguistic phenomenon in Arabic can be said to have started in the first Islamic century when Muslim exegetes and commentators set out to explain the contents of the Holy Qur?ān. Noticing that some of its vocabulary was, for one reason or another, indescribable in terms of being Arabic, they had to investigate those elements and account for their occurrence in that particular sector of the language, the Qur?ān, the paragon of their linguistic heritage.

However, linguistic borrowing as such was not studied in its own right. In fact the main concern of the Qur?ān commentators consisted in establishing whether or not the Qur?ānic vocabulary was wholly and purely Arabic, a task prompted not so much by linguistic interest or inquisitiveness as by a sense of religious duty. Greater importance was however attached to the phenomenon by later writers who, despite the fact that they did not dwell on the subject, became more and more interested in the tracing of foreign etymologies and the various aspects of the process involving their incorporation. For example, one of the earliest contributions made in this field was due to Sībawayh (d.770), who investigated certain phonological and morphological changes undergone by loanwords when assimilated into Arabic, thus revealing a more linguistically oriented approach to the problem than that of his predecessors (1361 AH, vol.2, pp. 18-19, 208-9, 375-6).

Succeeding Sībawayh, a number of other grammarians and philologists also contributed to the study of loanwords in some way or another and with varying degrees of depth and comprehensiveness. The contributions of these latter writers range from merely quoting Sībawayh's account, through citing new foreign borrowings, to stating personal attitudes towards the various issues of the problem or sometimes giving new accounts of already identified borrowings. Such attempts were made by al-Jāḥiḍh (d.868), Ibn Qutaybah (d.889), Ibn Durayd (d.933), Ibn Jinniyy (d.1002), al-Jawhariyy (d.?1005), al-Thaᶜālibiyy (d.1037), Ibn Sīdah (d.1066), etc.(7)

As has already been implied, none of these authors dealt with the question of borrowing as the exclusive subject of study. The first attempt of this kind came in the twelfth century with the appearance of "al-Muᶜarrab" by al-Jawāliqiyy (d.1144), which is wholly devoted to the early borrowings in Arabic with ample quotations to illustrate their occurrence, especially in poetry. Similar attempts were subsequently made by a number of other early writers whose works are also consulted in the following discussions.

In more recent times a large number of Arab writers have tried to tackle the problem as one aspect of the process of terminological expansion in modern scientific Arabic. We have done our best to investigate and review these writers' views and contributions which are scattered over a wide range of books, periodicals, bulletins, articles, etc.

4.3.2 The concept of Taᶜrīb and the status of borrowed vocabulary

Apart from the general designation *daxi:l* "non-native", as opposed to *?aṣi:l* "native, original", the foreign vocabulary of Arabic has traditionally been subjected to a further distinction. Before entering into details, it is useful to quote a relevant statement by al-Jawāliqiyy, with which he opens the introduction to his dictionary "al-Muᶜarrab" (p.51):

> This is a book in which we mention those foreign words which were spoken by the Arabs, uttered by the glorious Qur?ān, handed down in the traditions of the Messenger [Muḥammad] ... and His companions and followers, as well as those which the Arabs mentioned in their poetry and traditions.

Another statement of equal relevance is that made by al-Khafājiyy (d.1659) who, commenting on the previous one, adds: "What the later [Arabs] have arabicized is to be considered *muwallad*, which often occurs in books on philosophy and medicine".(8)

Thus, as can be inferred from these two statements, loanwords in Arabic are categorized into those that are *muᶜarraba:t*, as indicated by the title al-Jawāliqiyy uses for his dictionary (*al-muᶜarrab*), and those that are *muwallada:t* as specified by al-Khafājiyy. The former category represents those borrowings which were used during the Citation Periods (see pp.88-9); the latter, in subsequent times up to the present. The distinction may thus be said to have been drawn on a historical basis, i.e. to establish the particular time at which a given loanform has entered the language. This however is not the only consideration underlying the distinction. Whereas *muᶜarraba:t* are treated on an equal footing with native words, entered in dictionaries and used for further derivatives, *muwallada:t* are generally considered not so qualified as to be similarly treated; they are only latecomers and therefore not lucky enough to enjoy the good graces of their hosts. The argument put forward for this view is that unlike the latter, the former were used at a time when "correct" Arabic (*al-ᶜarabiyyatu l-fuṣḥa:*) was predominant, and that those who used them were the eloquent (*fuṣaḥa:?*) of the Arabs, who had a better command of the language in its classical form than later speakers. What seems to constitute the basis of this argument is the fact that the majority of classical borrowings were brought into line with the characteristic structure of native vocabulary, a phenomenon less frequently observed in the case of post-classical borrowings. In the light of the preceding statements it can easily be deduced that the main motive behind the distinction under discussion is to preserve the symmetry of the language and to check the overflow of new borrowings that might ultimately impair its character.

Thus it was often stressed by early scholars, as is the case in the present time, that loanwords should be made concordant with the phonological and morphological structure of Arabic. In a statement quoted by al-Suyūṭiyy, al-Jawhariyy defines the process of incorporating a non-native element as follows ("al-Muzhir", 1, p.268): "The arabicization of a foreign word consists in its being uttered by the Arabs according to their system." Another writer whose argument in support of the present principle can be described as the most definitive is al-Ḥarīriyy (d.1122) who, in a treatise on solecisms, declares that it is the Arabs' procedure that "if a foreign noun is arabicized, it is referred to what is parallel to it in their own language, [both] in type and measure".(9) He therefore marks as solecisms a number of borrowings on the grounds that their measures are not precisely Arabic. The following table contains a number of examples cited by him, both in their analogical and non-analogical forms, as well as a number of native words of analogical patterns as models:

Table 2. Analogical and non-analogical arabicization

Non-analogical		Analogical		Native word as a model
Loanform	Pattern	Loanform	Pattern	
ha:wan[10]	Fa:8aL	*ha:wu:n*	Fa:8u:L	*fa:ru:q*[11]
dastu:r[12]	Fa8Lu:L	*dustu:r*	Fu8Lu:L	*buhlu:l*[13]
sarda:b[14]	Fa8La:L	*sirda:b*	Fi8La:L	*sirba:l*[15]
barṭi:l[16]	Fa8Li:L	*birṭi:l*	Fi8Li:l	*sindi:d*[17]
šaṭranj[18]	Fa8LaLL	*šiṭranj*	Fi8LaLL	*jirdahl*[19]

Al-Ḥarīriyy's approach, then, holds that besides full phonological assimilation, arabicized words should assume morphological patterns fully identical with those of native ones. It is only then that they may be incorporated into the language. It should be mentioned though that to some others, post-classical borrowings, despite the fact that they may be fully naturalized, are still not to be considered as part of the Arabic vocabulary or be allowed a place in the dictionary since they were not used by the classical speakers. A case in point has already been mentioned on page 88: *an-niḥri:r* "intelligent man", which is relegated by al-Aṣmaᶜiyy to the category of *muwallada:t* and is described by him as not belonging to Arabic although it is fully naturalized and assumes a native pattern Fi8Li:L.

An important fact is that despite their unfavourable attitude towards non-analogical loanforms, neither al-Ḥarīriyy nor al-Jawhariyy were opposed to lexical borrowing as a principle, provided that what is borrowed is made consistent with the Arabic character. Thus loanforms which comply with the preceding measure are more often than not put on a par with native words. An often quoted saying of al-Fārisiyy, one of the most celebrated masters of the principle of analogy, is: "What is modelled [drawn by analogy] on the language of the Arabs is part of the language of the Arabs." Illustrations of the actual application of this statement are provided in 4.4.1 ii below.

Sībawayh's analysis of this phenomenon reveals a more moderate or rather less prescriptive approach than that of the above writers. He is more concerned with observing and analysing loanwords than laying down rules for their usage. He does not stipulate, for instance, that all foreign borrowings be fully assimilated into the analogical measures of Arabic. While recognizing that many such words do conform to this requirement, he does not restrict the concept of arabicization to analogical borrowing only. Rather, from his point of view, arabicization is a process whereby words of non-Arabic origin are taken over and used by the Arabs regardless of the fact that they may or may not be in full harmony with the native character of the language. Thus, examining the current vocabulary of his time, he tells us that the language contains both borrowings that have been assimilated on the analogy of native words, e.g. *bahraj* "false, fake" and *di:na:r* "a monetary unit" (on the analogy of *salhab* "having a long body" and *di:ma:s* "dungeon, etc." respectively), as well as others whose morphological patterns are not common to Arabic, e.g. *?ibri:sam* "silk" *?iF8i:laL and *?a:jurr* "baked brick" *Fa:8uLL. The term *muᶜarrab* as used by Sībawayh may thus be legitimately described as referring both to analogical as well as non-analogical arabicization (vol.2, p.375).

The above account concerning the attitude towards non-native elements and the phenomenon of lexical borrowing in general gives rise to a number of points which call for some consideration. The following is a discussion of these points and other relevant issues.

One main conclusion to be drawn is that there is a general tendency among Arab linguists to oppose the occurrence of non-native elements in their language or, when this is inevitable, to subject them to a process of full adaptation. It is also true that this tendency stems from the fear that the occurrence of foreign words might lead to the disruption of the characteristic structure of the language. While granting the honourability of these motives, one should not be blindfolded against certain elements of unrealistic thinking easily discernible in the arguments put forward by some writers.

To start with, there is no denying the fact that the speakers of any language are naturally inclined to use their own (native) words rather than borrowed elements, especially if the latter are not yet firmly established in their system. However, it is also undeniable that it is a natural phenomenon among languages to borrow from one another, a phenomenon which may be accounted for by several factors. For one thing, it may safely be argued that none of the languages of the world could be said to contain human thought in all its totality and diversity. The same argument also holds true with regard to the culture of a particular nation. In other words, neither languages nor cultures are sufficient in themselves. That languages borrow from each other is only a reflection of the tendency of their respective speakers to adjust themselves in a give-and-take sort of cultural interaction. While a particular language may be described as being the mirror of the culture or society it represents, it may not necessarily be equally capable of reflecting the picture of other cultures and with the same degree of adequacy and detail. It follows then that lexical borrowing is a natural consequence of cultural inter-influence between nations.(20)

Opposition to borrowing is often based on the argument that Arabic need not resort to external resources for meeting its new lexical needs since it has a vast linguistic heritage at its disposal. A verse often quoted to this effect is from Ḥāfiḍh Ibrāhīm in which, speaking on behalf of the Arabic language, he says:

> *?ana: l-baḥru fi: ?aḥša:?ihi d-durru ka:minun*
> *fa-hal sa:?alu l-ġawwa:ṣa ᶜan ṣadafa:ti: ?*
> I [the language] am the sea in whose depths pearls [i.e. words] are hidden,
> have they [those who look for new ones] ever asked the diver about my oysters?

That Arabic has a rich vocabulary and that it rests upon a rich cultural basis cannot be denied,(21) but this need not imply that it encompasses all human thought. Those who engage in translation, no matter how adequate they are, soon come to the realization that the scope of thought often extends beyond that of language, which accounts for the difficulty they sometimes meet in translating certain words from one language into another (see Ḥassān, p.26).

An important fact which some purists do not seem to have taken into account is that of all the domains in which linguistic interference may take place, lexical borrowing has the least possible influence (if any) on the characteristic structure of the borrowing language. Whereas the borrowed vocabulary could be enormous, new structural features due to borrowing from another language are very rare. These are accepted only when they correspond to the developmental tendencies of the language in question (see Weinreich, p.25). Accounting for this situation, Haugen (p.224) states that structural features are "correspondences which are frequently repeated ... they are established in early childhood, whereas items of vocabulary are gradually added to in later years. This is a matter of the fundamental patterning of language: the more habitual and subconscious a feature of a language is, the harder it will be to change." Thus whereas structural features are rarely imported from one language into another, lexical importation is a widely observed phenomenon in all languages. It is a common observation among modern linguists that the vocabulary of a language is the domain most penetrable to interference from other languages.(22)

Another fact to be taken into consideration is that lexical borrowing does not always seem to be a matter of whether or not the language involved is self-sufficient. The process may be motivated by a variety of other factors. Foreign words may thus be borrowed not because there is a real need for them but because they are intended to produce a certain effect, e.g. to indicate an aspect of culture or an environment which is not typically characteristic of the native society. A case in point is the use in some Arab countries, Iraq for one, of the indigenous and traditionally established name *maqha:* for the place where tea, coffee and sometimes soft drinks are bought or drunk, whereas *ba:r* "bar" and *ka:zi:no:* "casino" refer to public buildings where the atmosphere is rather suggestive of an external (Western) influence, e.g. alcohol-drinking, dancing, gambling, etc. Another interesting example may be cited from Ḥasāwi Arabic, where the office of an Arab merchant is designated by the native word *maktab* while that of an American by the loanform *ḥafi:z* "office" (Smeaton, p.72).

It is also true that sometimes words are borrowed for their snob value. The borrower in this case may feel that using non-native words confers on him an air of learnedness, and so distinguishes him from other members of his community (cf. Ullmann, p.133).

There is sufficient evidence that a foreign word may for one reason or another find its way into Arabic despite the existence in the latter of sometimes more than one native equivalent. This phenomenon can be traced to very early stages of the language. A case in point is the current word for "aubergine", namely *ba:ḏinja:n*, which Arabic has adopted from Persian ba:dinga:n. Speakers of present-day Arabic might be surprised to know that their ancestors had a multiplicity of native names for the same vegetable, e.g. *al-ḥadaq*, *al-maġd*, *al-qahqab*, etc. (Smeaton, p.72).

A relevant phenomenon increasingly noticeable in modern Standard Arabic is that a foreign term is resorted to as a result of the fact that different writers may vary widely in the type of rendition they use for one and the same term; besides, some of these renditions are sometimes seen as lacking in adequacy and precision. For instance, the linguistic term "morpheme" has been given the following renditions by a number of Arab linguists: *ᶜa:milu n-nasaq*, *ᶜa:milu n-nisbah*, *ᶜa:milu ṣ-ṣi:ġah*, *dawa:llu n-nisbah* and *lafḏu l-ʔirtiba:ṭ* (see Abu al-Faraj, fn.3 p.82). However, due to this inconsistency, other writers would rather use the loanform *ma:rfi:m* instead of any of the preceding renditions (see e.g. al-Ṣāfi, p.177).

Euphemism is another factor at work in favouring the adoption of a non-native word. By virtue of being non-native and hence probably not so familiar, the foreign borrowing may sound less offensive than its native equivalent which is a taboo word. The loanform *tuwa:le:t* "water-closet" from French toilette, for instance, is often used instead of native words. We must hasten to mention though that in this particular case the same loanform may itself soon come to be replaced, as it becomes more and more associated with the referent and hence gradually loses its pleasantness.

Some purists seem to be of the mistaken view that borrowing is a belittling feature in language, that it is a sign of inadequacy and imperfection. We have already said that borrowing is a common phenomenon among all human languages. A borrowing language may itself be drawn on by several others, and as far as Arabic is concerned it has been mentioned above that a large number of Persian words trace their origin back to it. We shall also see below that many Arabic words have found their way into several other languages. Contrary to the above-mentioned view, it could thus be argued that the fact that a language is capable of making use of this method as one of the means whereby it can meet its urgent needs is an indication of its capacity to survive, evolve and meet the ever-growing demands of modern civilization. Besides, in the case of Arabic (and this may also be true of other languages), borrowing can be expected only as a last resort, i.e. after all other methods of lexical creation have failed to produce an appropriate native equivalent; thus there need be no fear of these borrowings overwhelming the language. Furthermore, the fact that the words Arabic borrows are often made consistent with its characteristic structure is another safeguard against what may be considered a threat to its inherent qualities.

4.3.3 Post-classical writers and the systematization of Taᶜrīb

It will have been gathered from the preceding discussions that there was a general tendency on the part of early Arabs to control and regulate the process involving the introduction into their language of words belonging to other languages. Their efforts in this respect consist in concluding a number of facts concerning the phonological and morphological changes which borrowed words undergo as a result of being assimilated into the lexical system of Arabic. In addition to analogical arabicization, a detailed discussion of which is provided in 4.4 below, two main issues can be distinguished as constituting the attempts carried out in this respect:

1 the identification of loanforms
2 the assimilation of non-native sounds

The following discussions are intended to give a detailed account of these attempts, the results they have yielded and their potential for the systematization of the process in question. The significance of investigating and evaluating these issues stems from the fact that Arab language academies, as well as the majority of other concerned bodies, have always recommended that the rules and principles established by early Arabs be the basis of any further attempt to investigate the role of borrowing as one of the means of lexical expansion in present-day Arabic. The crucial point is that many writers have tried to give the impression that the early accounts of the issues under discussion have been fully adequate and exhaustive. While granting the high creditability of early achievements, we shall see however that the state of affairs is not precisely as has been described. It should be mentioned though that this is largely due to the nature of the problem rather than to the efforts so far expended in its investigation.

4.3.3.1 Loanform identification

One aspect of the efforts made by early Arabs to investigate loanforms in their language is their use as criteria for distinguishing such forms of a number of facts concerning the types of phonological and morphological patterning characteristic of their language and the differences revealed by the items of vocabulary borrowed from other languages.

A. Sound patterns as a criterion for identifying loanforms

One of the creditable achievements of early Arabs in this respect if their observance of the fact that the combination of sounds in their language is not a random process but rather these combinations are governed by certain restrictions concerning which they did not fail to make some very significant statements. Awareness of the types of sound combinations that may or may not be permissible in their language enabled early investigators to designate as non-native formations those words whose phonological structure is not in keeping with what is characteristic of their own formations. Before illustrating this, we may as well acquaint ourselves with a number of conclusions established in this respect.

In his account of Arabic consonant combinations, the phonetician and grammarian al-Khalīl bin Aḥmad (d.786) established that "when three letters of the alphabet are articulated from one place, they do not combine together as radicals in one word, e.g. b f m as labials can never be found together in an Arabic word. The same applies with ṯ ḏ ḏ̣ as dentals and with ḥ ᶜ x as gutterals" (quoted by Darwīsh, vol.1, p.3). We may thus designate as a least accepted consonant combination in Arabic the one which involves the use of homorganic radicals in the same word.

Turning to what may be said to represent the most typical or most acceptable to Arabic, we may quote al-Jawāliqiyy's statement: "As regards Arabic formations, the most excellent are those formed by letters which are distant in their place of articulation."(23) This statement has been described by Greenberg as the closest to a general formulation he has been able to discover in his own study of "The Patterning of Root Morphemes in Semitic".(24)

Another praiseworthy contribution to the study of Arabic sound patterns is that of Ibn Jinniyy (d.1002), whose investigation is characterized by a remarkable degree of validity and exhaustiveness. The information contained in his "Sirr Ṣinā^{c}at al-I^{c}rab"(25) is most revealing, a fact substantially supported by the close correspondence we have noticed between his findings and conclusions and those arrived at by Greenberg in his aforementioned study. Ibn Jinniyy's other book, "al-Khaṣā?iṣ", also contains some information pertaining to the phonetics and phonology of Arabic.

Scattered references to various aspects of this topic may also be found in the writings of other Arab writers whose works will be cited in the course of our discussion.

In the light of the studies mentioned above, it is possible to identify a number of features as characteristic of the patterning of Arabic sounds. The identification of these features is seen as necessary, as it enables us to achieve the following objectives:

(a) Establish a set of criteria on which to determine whether or not or to what extent a certain loanform conforms to the phonological structure of Arabic.

(b) Determine the type and nature of change a foreign word may be expected to undergo as a result of its being assimilated into the phonic structure of Arabic.

(c) Assess the impact of the borrowing process on the standard norms of Arabic sound combinations, e.g. whether non-native patterns have been incorporated or have constituted a novel feature in the language at its present stage.

It has already been mentioned that the overwhelming majority of Arabic linguistic forms are based on a triconsonantal morpheme or root. Unless otherwise stated, this root will therefore be employed as our point of reference. We shall use the symbols C1 C2 C3 to indicate respectively the initial, medial and final consonants constituting any triconsonantal root in Arabic. The positions of these consonants will be indicated by the Roman numerals I II III respectively. When used without a number, the C stands for any consonant in the position occupied by it. The following are the characteristic features referred to above:

(i) We have said that Arabic does not permit the occurrence of three homorganic consonants in the same root. It is also true that two homorganic consonants do not coexist in positions I and II in the same root. This applies in the case of identical consonants too. Thus there are no combinations beginning with *b-m-C, *ṯ-ḏ-C, *n-ṅ-C, etc. According to Greenberg (p.162) this characteristic is also true of other Semitic languages.

(ii) Homorganic consonants are again almost entirely excluded in positions II-III. Sequences such as *C-b-m, *C-t-ṭ, *C-s-z,

*C-k-j, *C-ᶜ-ḥ, *C-?-h, etc. are extremely rare in Arabic. A form like *šabim* "cold", where C2 and C3 are both labials, is an instance of such rare occurrence.

(iii) With regard to the occurrence of two identical consonants in II-III, this is quite frequent. Thus while a sequence like *k-k-f is non-existent, it is quite common to have k-f-f "to give up", f-k-k "to disjoin", d-r-r "to flow copiously", r-d-d "to return", and so on.

The present pattern, which constitutes the well known geminite sub-type of Arabic verbs, also characterizes a large number of nouns. There are for instance *qadd* "shape, figure", *sinn* "tooth", *lubb* "core", etc.

(iv) Positions I-III. According to Greenberg's (p.162) statistical investigation concerning these two positions, "there is marked, but less rigorous exclusion of homorganic, including identical consonants than in other combinations of positions". Ibn Jinny, likewise, has established that concurrences of identical consonants as C1 and C3 are less frequent than those where the two consonants occur in positions II-III (see Bakalla, p.273). Here are some examples of Arabic words containing identical first and third consonants:

natin	putrid
salis	docile, flexible
daᶜd	a girl's name
qalaq	anxiety, worry
yaday-	to touch or hit (somebody) on the hand (*yad*)

(v) The occurrence of homorganic consonants in positions I-III is subject to the following restrictions. While there are a few cases involving /b/ as C1 and /m/ as C3 (as in *barim* "disgusted", *basam-* "to smile", etc.), combinations where C1 is /m/ and C3 /b/ are entirely excluded. Likewise /d/ as first consonant never occurs with /d/ or /t/ as third, nor vice versa. On the other hand, if we take the group of homorganic consonants consisting of /l/, /r/ and /n/, we find that Arabic has numerous roots where /r/ and /l/ occur in the first and third positions respectively. There are for example r-k-l "to kick", r-s-l "to send", r-ḏ-l "to be mean", etc. Again, roots with initial /r/ and final /n/ are quite common, e.g. r-k-n "to lean upon", r-h-n "to pawn", r-s-n "to tie with a rope" etc. Examples where /n/ comes first, and /r/ third, are also frequent, e.g. n-ḥ-r "to slaughter", n-ṣ-r "to assist", n-ṯ-r "to scatter", etc. However instances of /l/ and /r/ as C1 and C3 respectively are extremely rare. The same also applies to /k/, /x/ and /g/, regarding their incapability of co-existing in positions I-III in the same root.

What has so far been said concerning Arabic sound combinations is intended to introduce the reader to the nature and types of phonological change foreign words may be expected to undergo on passing into Arabic, and more specifically when fully integrated into the native system. Additional information is given below whenever the occasion arises. The reader may also refer to the account of Arabic syllabic structure and consonant clusters in 1.4 above.

To recapitulate, the above facts concerning the patterning of
Arabic sounds were used by early investigators as strictly linguistic criteria for identifying the non-native elements in their language. Their attempts in this respect consist in analysing the
phonemic constitution of these elements and pointing out the points
of difference or divergence they reveal when examined against native
patterns.

Thus, as al-Zabīdiyy (d.997), for instance, tells us the
form saḏa:b (probably from Persian suda:b "a rue plant") is not
genuine Arabic because /s/ and /ḏ/ never coexist in the same
Arabic root (vol.1, p.295). Ibn Jinniyy also points out that due
to the inadmissibility in Arabic of a syllabic structure of the
type CVCC, which is common in Persian, e.g. ma:st "yogurt", this
has been given the Arabic rendition mast where the structure CVCC
is acceptable (see Bakalla, p.382). The following are a number of
other sound combinations which have been designated as not characteristic of Arabic and are largely restricted to early borrowings:

1 /n/ + /r/ e.g. narjis "narcissus" (Pah. nargis/Gr. νάρχισσος)

2 /d/ + /z/ e.g. muhandiz "geometrician"(26) (a derivative of
   Per. anda:za "measure, dimension")

3 /j/ and /q/ in the same form e.g. manjani:q "mangonel"
   (Gr. μαγγανιχόν)

4 /s/ and /j/ in the same form e.g. ṣanj "cymbal" (Per. čang)

5 Quadri- and quinquiconsonantal forms which do not include one
   or more of the following sounds: /b/ /f/ /m/ /n/ /r/ /l/. This
   observation involves an often stressed characteristic of Arabic
   lexical formations, and therefore requires some special consideration.

Early Arab phoneticians describe these consonants as the "smoothest
of Arabic sounds" (ʔaxaffu l-ḥuru:f). In his account of these consonants, al-Khalīl divides them into two categories, on the basis
of their point of articulation (vol.1, p.57). Thus the series
/b,f,m/ are described as labials (šafawiyyah), while /n,r,l/ as
liquids (ḏalaqiyyah, produced by the tip of the tongue, ḏalaq).
Hence the term al-ḥuru:fu ḏ-ḏalaqiyyah or ḥuru:fu ḏ-ḏala:qah which
is commonly used nowadays to cover both categories in the sense
that smoothness is an inherent characteristic in them.(27)

According to Ibn Jinniyy al-ḥuru:fu ḏ-ḏalaqiyyah involve some
"interesting secret" (sirrun ṭari:f), which is that any quadriliteral or quinquiliteral noun in Arabic must have at least one of
these consonants, otherwise it is non-Arabic, i.e. borrowed.(28)
More recently, a modern scholar, Ibrāhīm Anīs, has proved statis-
tically that the sounds /l,m,n/ have the highest frequency of
occurrence of all the sounds in Arabic. In each 1000 sounds, /l/
occurs 127 times, /m/ 124 times, and /n/ 117 times (Anīs, 1950,
p.178).

It should be mentioned however that as far as the distinction
between native and non-native forms is concerned, this criterion
alone is not always valid; for one or more of the phonemes in

question may already exist in the foreign model, in which case the fully integrated form would not betray its foreignness. Take for instance the word *faylasu:f* "philosopher" from Gr. φιλόσοφος, where the labial /f/ and the liquid /l/ in the Arabic rendition are represented by more or less equivalent sounds in the original model. A more recent example, *tilifo:n* "telephone", contains three of the *d̲ala:qah* consonants, namely /l,f,n/. It should be pointed out though that in the latter case, unlike the former, the non-nativeness of *tilifo:n* is still evidenced by the fact that it does not correspond to a native morphological pattern.(29)

The point raised in the preceding paragraph paves the way for us to discuss a relevant problem involved in the identification of loanforms and which early scholars have not always handled with adequate caution. We have already seen some cases (and more detailed instances are yet to follow) of lexical borrowings whose adaptation to Arabic has been such that one can hardly find a clue to their foreign descent, i.e. on the basis of their phonological and morphological structure only. The fact that such forms give birth to further derivations like any other native words (see 4.4) serves as another factor enforcing a decision in favour of considering them of Arabic origin. Fortuitous similarity between a loanform, especially when fully integrated, and another form of native origin is another important factor accounting for the inaccuracy involved in establishing the origin of some borrowings. In view of these facts and in the absence of adequate etymological investigation, fully integrated loanforms have sometimes been mistakenly thought to be originally Arabic. A case in point is the loanform *?iqli:d* "key"(Gr. χλειδα), which is claimed to be kindred with native forms derived from the root q-l-d "to adorn with a necklace". Another case is represented by *?asṭurla:b* "astrolabe" which has been described as being a native formation consisting of two elements, *?asṭur* "lines" and *la:b* "name of a town".(30) In actual fact, however, this word traces its origin back to Gr. αστρόλαβος. Sometimes a loanform is related to a native root although its non-native origin is already established. An interesting case in point is the Qur?ānic word *ṣira:ṭ* "path, way". While admitting that it derives from Latin strāta, Luwīs Maᶜlūf (1966, p.330) enters this word under the Arabic root s-r-ṭ "to swallow" and defines it as "a clear path, for its pursuer disappears like swallowed food".

B. Morphological patterns as a criterion for identifying loanforms
The sounds constituting a loanform may all be familiar to Arabic, the sequences in which they appear may also reveal no deviation from current combinations, yet such a form may be recognized as originally foreign on the basis of the type of morphological pattern it has. This principle has been employed by early scholars as one of their criteria for distinguishing non-native items of their vocabulary. Relevant instances of the application of this criterion have already been included in our discussion of the issues raised in 4.3.2 above (for further examples see Sībawayh, vol.2, pp.375-6).

The attention of early investigators was mainly concentrated on the two criteria given above, i.e. sound combinations and morphological patterns, which may be attributed to the significant role these two principles play in the characterization of native vocabulary. However, we find it useful to include the following as further criteria.

C. Incongruity of loanforms with native roots
The non-native identity of a given loanform may also be inferred, to some extent, from the fact that there is no root in the language which may be said to underly it both from the point of view of the meaning it conveys and the radical morpheme constituting its source of derivation. For instance, if we take forms like *fa:nu:s* "lantern" (Gr. φανός), *busta:n* "garden" (Per. bu:-sta:n "place of perfume") and the more recent borrowings *ra:da:r* "radar", *film* "film",(31) etc., we find that they are not traceable to native origins, for Arabic has no such roots in its native stock as *f-n-s, *b-s-t-n, *r-d-r or *f-l-m, respectively. This however need not imply that such roots may not later on be abstracted from their corresponding loanforms and even used for further derivation, thus establishing themselves as operative items of the current vocabulary. Indeed some of the foregoing examples have already acquired such a status; among the currently used derivatives of b-s-t-n are *bastanah* "gardening, horticulture" and *busta:niyy* "gardener", and *film* has been given the analogical plural form *?afla:m*, like *ṭifl* "child", *?aṭfa:l*"children".

D. Stress patterns as a criterion for identifying loanforms
Early investigators made no mention whatsoever of CA stress patterns as a criterion for the identification of non-native forms. The reason for this seems to have been the fact that all loanforms have been integrated in full conformity with the native stress system of the language (see 1.5). This observation is also true of loanforms borrowed in modern times. Here are some examples of early loanforms stressed in accordance with the Arabic system (native examples of corresponding syllabic structure are also provided:

Loanform		Native Word	
fayla'su:f	philosopher	*ḥayza'bu:n*	old hag
?asṭur'la:b	astrolabe	*?istiq'la:l*	independence
fir'daws	paradise	*jil'lawḏ*	stout camel
fi'lizz	metal	*xi'ḍamm*	sea
di'maqs	raw silk	*hi'zabr*	lion
qusṭanṭi:'niyyah	Constantinople	*?istiqra:'?iyyah*	inductiveness

E. Initial clustering as a criterion for identifying loanforms
It has been said above that SA does not admit a consonant cluster at the beginning of a word. Items of the vocabulary where such clusters do occur may thus be easily identified as non-native elements. Instances of such cases are not difficult to find among recent borrowings, especially in the language of science, as we shall see. It should be noted, however, that this is only the case where the loanforms involved are not fully naturalized; otherwise, if they are subjected to analogical arabicization, they are no

longer characterized by this feature. As far as early borrowings are concerned, we have noticed that initial combinations of consonants are declusterized in most, if not all, cases. This is done either by inserting a vowel after the first consonant or, which is more frequently the case, though the use of a prosthetic syllable of the type CV, where C is a glottal stop: /?/, and V a short vowel: /a/, /i/ or /u/. In the latter case, the first consonant in the cluster together with the prosthetic syllable preceding it form a new syllable of the type CVC, which is acceptable to Arabic, Here are some examples of early cases to illustrate the point:

balġam	phlegm	Gr. φλέγμα
dirham	monetary coin	Gr. δραχμή
?afla:ṭu:n	Plato	Gr. πλάτων
?isṭabl	stable (n.)	L. stabulum
?isḟanj	sponge	Gr. σπόγγος
?iqli:m	climate, region	Gr. κλῖμα (32)
?uṣṭu:l	fleet	Gr. στόλος

4.3.3.2 Assimilation of non-native sounds

Another aspect of the contribution of early Arabs to the investigation of loanforms in their language is their attempt to study and account for the sound changes involved in the process of adopting these forms. We have seen above how some (but not all) of the early scholars are adamant on the question of adopting foreign borrowings to the morphological patterns of the language. Regarding the assimilation or replacement of non-native sounds, i.e. those which do not constitute part of the native phonological system, their opinion, which is no less adamant, is unanimous. The process, as they see it, is not merely a matter of linguistic tendency on the part of speakers to substitute native sounds for foreign ones, but rather a deliberate procedure whereby they consciously avoid what might constitute an alien feature in their language. According to al-Jawāliqiyy (p.54), for instance, "the substitution [of foreign sounds] is conditional lest they [the Arabs] should include in their speech what is not part of their [inventory of] sounds". The occurrence of a foreign sound in an Arabic or arabicized word is always regarded as a factor of malformation which deprives that word of its elegance and he who uses it would run the risk of being criticized as incapable of speaking proper Arabic. It is no wonder then that there is hardly any mention in the early literature of foreign sounds being retained or incorporated into the language.

One conclusion to be drawn from the early attempts at studying sound changes is that they were based on a rather limited range of instances, which accounts for the lack of exhaustiveness perceived on the results they have yielded. They were also largely confined to Persian borrowings, probably because Persian was the foreign language with which early investigators were best acquainted. Let us examine a representative sample of these attempts.

According to al-Suyūṭiyy's authorities ("al-Muzhir", 1, p.274) ten of the Arabic phonemes are involved in the process of phonological assimilation of loanforms. These are divided into two groups:

(a) those that are regularly used as substitutes for foreign sounds that do not exist in Arabic and include /k, j, q, b, f/; and
(b) those that are substituted, but not regularly, for certain sounds contained in borrowed forms but which are also part of the Arabic system. These include /s, s, ᶜ, l, z/ (see al-Suyūṭiyy, "al-Muzhir", 1, p.274).

As far as group (a) is concerned, we are told that two or more of these sounds may be substituted for one and the same foreign counterpart in different words, sometimes even in different versions of the same word. For instance:

Per. /p/ as in	parand	→ Ar.	*birind/firind*	sword[33]
	parga:r	/k/	*birka:r*	a compass[34]
Per. /g/ as in	gora:b	→ Ar. /j/ as in	*jawrab*	stocking[35]
	gardan	/q/	*qard*	neck[36]

In addition to /p/ and /g/ other non-native sounds contained in the original forms of early borrowings include /c/, /z/ and /v/. Here are some examples to illustrate the way these sounds tend to be treated in Arabic:(37)

	/ṣ/ e.g.	*čak*	→ *ṣakk*	legal document, cheque[38]
č →	/š/ e.g.	*ču:bak*	→ *šawbak*	a rolling pin[39]
	/j/ e.g.	*čang*	→ *junk*	a harp, lute[40]
ž →	/z/ e.g.	*ži:wa*	→ *zi?baq*	mercury[41]
		kaž	→ *qazz*	coarse silk[42]
v →	/w/	*asva:r*	→ *?uswa:r* or *?iswa:r*	a horseman[43]

With regard to group (b) sounds, i.e. /s, š, ᶜ, l, z/, it has been mentioned that these are sometimes used to replace other sounds contained in borrowed forms, but which also exist in Arabic. For example, in a loanform like Per. anda:za (p.105), the sound /z/, although one of the native phonemes of Arabic, is not retained in the arabicized version of the word, but is replaced by another native sound, /s/, as in *handasah* "geometry", *muhandis* "geometrician, engineer". Such changes are largely accounted for by the fact that they are motivated by the general tendency of Arabic sounds to combine in certain sequences rather than in others, a tendency which is also extended to arabicized forms by way of bringing them into line with the characteristic structure of native vocabulary.

It should be added that a wider investigation of early borrowings would reveal a number of other native sounds which are used in the same way as those included in group (b) above. Some of these sounds belong to group (a). Here are some examples of early Persian borrowings:

/q/ for /k/ e.g.	karwa:n →	*qayrawa:n*	a caravan
/q/ for /z/ e.g.	a:b-rez →	*?ibri:q*	water jug
/q/ for /x/ e.g.	dih-xa:n →	*dihqa:n*	a leading personality
/j/ for /k/ e.g.	ku:šk →	*jawsaq*	a palace
/f/ for /b/ e.g.	xoš-a:b →	*xiša:f/xuša:f*	sherbet of raisin juice

A significant phenomenon to be mentioned here is the widely observed tendency of early speakers, especially translators, to replace the sounds /t, d, s, k/, which occur in borrowed words and which also occur in Arabic, by their velarized counterparts /ṭ, ḍ, ṣ, q/ respectively. A relatively early reference to this phenomenon is made by Blochet, who, without accounting for it, states that "En général, dans les transcriptions des mots étrangers, l'arabe emploie toujours ses consonnes emphatiques". Here are some of his examples:

ʔiṣṭaxar	<	Per. Ixtaxar (name of a town)
ʔarisṭa:ṭa:li:s	<	Gr. 'αριστοτέλης Aristotle
qusṭanṭi:niyyah	<	Constantinople
baṭṭa:riyyah	<	Fr. batterie (Blochet, p.270)

An explanation for this phenomenon seems to be that (as will have already been gathered) early Arab translators were keen on rendering loanforms as consistent with the character of their language as possible, a tendency largely attributable to their desire to preserve the distinct character of Arabic and maintain its characteristic features that distinguish it from other languages. Emphatic sounds, being among the most distinguishing features in this respect, must thus have been felt to be more capable of embodying this distinction than the non-emphatic ones which are common to most languages.

The same tendency is also observable in the works of other more recent translators, especially those with a classical education. Among the most representative is al-Ṭahṭāwiyy (1801-73), in whose account of European civilization and the early cultural encounter between modern Europe and the Arab world we find a large number of loanforms characterized by this phenomenon. We shall concentrate on volume 2 of his work in which he relates various aspects of his visit to Paris in 1826. The significance of examining the non-native forms introduced by this author is that he is the first Arab writer of modern times to come face to face with the problem of dealing with a large number of loanforms, most of which had not been used in Arabic before. The picture may thus be said to resemble the state of affairs in the first renaissance when Arabic had its earliest encounters with Greek scientific borrowings.

Al-Ṭahṭāwiyy's borrowings reveal a strong tendency to follow the early Arabs' example by substituting emphatic consonants for non-emphatic ones. Here is a sample of his borrowings from French:

1	*qalqu:ṭa:*	<	Calcutta	(p. 28)
2	*al-la:ṭi:niyyah*	<	Latin (language)	(p. 95)
3	*nu:yurg*	<	New York	(p. 28)
4	*al-mi:ka:ni:qa:*	<	mecanique	(p. 22)
5	*al-qa:ṯu:li:qiyyah*	<	catholicisme	(p. 96)
6	*al-ʔinisṭi:ṭu:t*	<	l'institut	(p.163)
7	*ar-ra:sṭra:ṭu:ra:t*	<	restaurants	(p.123)

It should be noted however that many of al-Ṭahṭāwiyy's borrowings have later been either re-borrowed or replaced by native equivalents. The forms just quoted, for instance, are nowadays given the following renditions:

1 *kalkatta:*
2 *al-la:ti:niyyah*
3 *niyu:yu:rk*
4 *al-mi:ka:ni:ka*
5 *al-ka:ṯu:li:kiyyah*
6 *al-maᶜhad*
7 *al-maṭa:ᶜim*

As can be easily deduced from these latter renditions, present-day speakers of Arabic do not reveal as much of a tendency towards the substitution of emphatic consonants for non-emphatic ones (i.e. in recent borrowings). (However, instances exhibiting the older tendency are not entirely lacking in the present time.) An explanation for this development may be that acquaintance with foreign languages and the way their words are pronounced is far more widespread nowadays than ever before. At least as far as written language is concerned, newly introduced borrowings (and sometimes even their native equivalents) are more often than not accompanied by their original foreign models, which makes it easy for the reader to familiarize himself with the actual sounds constituting the word underlying the loanform.

It was hinted earlier that a lack of precision is discernible in the accounts given by early investigators. This unfortunately is also true of accounts given by more recent scholars. For instance the fact is often forgotten that early Muslim contacts with Persia were with a people speaking Middle and not Modern Persian (Jeffery, p.15). Ignorance of this fact has resulted in these writers giving erroneous accounts of many Pahlavi borrowings which are mistakenly thought to belong to later Persian. According to Addi Shirr (p.4), for instance, "[early Arabs] added a /j/ or a /q/ at the end of arabicized words, especially those ending in /h/. Thus, they said *jawzi:naq* or *jawzi:naj* for gawzi:nah, and *qurbaj* or *qurbaq* for kurbah". The same view is held by many other writers. This however is not precisely the case. The explanation is simply that the two borrowings, ġawzi:nah "a sweetmeat made of walnut" and kurbah "a shop", do not in fact represent the actual models of the Arabic renditions given above. These Persian words are themselves reflexes of older forms from an earlier stage of the language, i.e. Pahlavi or Middle Persian, where they are terminated by the sound /g/ or /k/ (see Blochet, pp.266-7). In other words, Modern Persian /h/ represents the Pahlavi suffix /-ag/ or /-ak/. It thus follows that the sound /j/ at the end of the Arabic version is, contrary to Shirr's claim, not an addition, but rather a substitute for the Pahlavi suffix /g/ since it does not constitute part of the Arabic phonological system. We have already seen some examples where this sound is similarly treated in other borrowings and sometimes replaced by /q/ (p.109). With regard to the suffix /-ak/, again mention has been made of the fact that the use of Ar. emphatic /q/ to replace /k/ in early borrowings was a widespread phenomenon (p.109).

Evidence in support of our argument may be derived from the fact that almost all other cases where a loanform is claimed to have as its model a Persian word ending in /-ah/ are terminated in Arabic by /j/, /q/ or /k/ and not other consonants, although there seems to be no other reason why this should be the case. Examples:

ṭa:zaj	fresh	Mod.Per. ta:zah
sa:ḏaj	simple, plain	Mod.Per. sa:dah
?istabraq	brocade	Mod.Per. ?istabrah
baydaq	chess piece	Mod.Per. piya:dah

4.4 Analogical arabicization: its significance in the light of present terminological needs

The term "analogical arabicization", as will have already been understood, refers to the process whereby a loanform is rendered fully harmonious with the Arabic character, from the point of view both of its phonological and morphological structure. In other words, the result of this process should be describable in terms of being a lexical item whose constituent phonemes form part of the native phonological system, whose morphological pattern is exhibited by native words and which can be said to derive from a root consisting of three or four radicals (which are abstracted from the original loanword). Arab language academies as well as the great majority of writers, both early and modern, attach great importance to this process as a necessary procedure for regulating the transfer of loanforms into Arabic. However, various important aspects of this process, especially its practicability in the language of science, have not yet been seriously studied and therefore call for discussion. In the following pages we shall try to put forward a number of arguments which may be said to justify the adoption of this process, as well as others which may be seen as factors limiting its applicability in certain cases.

4.4.1 Arguments for analogical arabicization of loanforms

(i) The modification of loanforms in accordance with the predominant character of the borrowing language can be said to represent a linguistic phenomenon common to all languages, the only difference being that the nature and degree of modification varies from one language to another depending on the typical character of the borrowing language in question. Let us take English as one case in point and see how it has treated one of its non-native elements, namely "alkali" from Arabic *al-qily* "the ashes of the saltwort plant" (Webster, p.29).

To start with, the Arabic definite article *al-* (which always appears in conjunction with the following noun) has been rendered as a constituent part in the English replica of the Arabic model, thus "alkali". This however does not mean that the latter is treated as a definite noun in English. Like any other noun it may be made definite or indefinite through the use of the English article "the" or "an". Compare also the way English "alkali" and Arabic *al-qily* have been employed for further derivation:

alkaline	*qilwiyy*
alkalinization	*taqliyah*
alkaloid	*šibhu qilwiyy*
alkalosis	*qula:?*
alkimeter	*miqya:su l-qilwiyyah*

And just as Arabic follows its own system in pluralizing its non-native forms, so do other languages when they pluralize Arabic borrowings. The situation may be illustrated by examples of mutual borrowing between Arabic and English/French:

(a) English/French loanforms as pluralized in Arabic:

Singular		Plural	
ka:mirah	camera	*ka:mira:t*	(by suff. *-a:t*)
landaniyy	Londoner	*landaniyyu:n*	(by suff. *-u:n*)
malyu:n	million	*mala:yi:n*	(by intervocalic change Fa8a:Li:1)
bank	bank	*bunu:k*	(by intervocalic change Fu8u:L)
mi:l	mile	*?amya:l*	(according to the pattern ?aF8a:L)

(b) Arabic loanforms as pluralized in English:

magazine	*(maxzan)*	magazines	*(maxa:zin)*
camel	*(jamal)*	camels	*(jima:l)*
giraffe	*(zara:fah)*	giraffes	*(zara:?if, zura:fa:)*
minaret	*(mana:rah)*	minarets	*(mana:?ir)*

Sometimes an Arabic word is so drastically modified through the process of borrowing that it looks as though it belongs to another language. A case in point is the star-name *rijlu l-?asad* "lion's paw" which is given a very Latin appearance through the rendition "Regulus" (i.e. in Latin) (see Taylor, p.570). Many such cases, where Arabic originals are rendered into hardly recognizable mis-transcriptions can be found in the medieval Latin translation of Ibn Sīna's (Avicenna's) great medical encyclopaedia, "Al-Qānūn", (The Canon). Here are some examples:

Arabic:		Latin:
al-ᶜuṣᶜuṣ	coccyx	alhosos
al-qaṭan	the lumbar region	alchatim
al-ᶜajuz	the sacrum	alhauis
an-nawa:jiḏ	the wisdom teeth	nuaged/neguegidi

Compare also:

Eng. admiral, Fr. amiral < Ar. *?ami:ru l-baḥr* "prince of the sea"
Eng./Fr. arsenal < Ar. *da:ru ṣ-ṣina:ᶜah* "house of manufacture"
Eng. benzoin, Fr. benjoin < Ar. *luba:n ja:wiyy* "frankincense of Java"

(ii) An important argument in favour of analogical arabicization is that it preserves the symmetry of the language and is an economical process. With regard to various aspects of its structure, Arabic has frequently been described as being radically different from many of the well-known languages of Europe (see Ferguson, p.183). One of the most celebrated of these aspects is the system of analogical derivation which constitutes the basic foundation of the language and its most productive method of word formation. The incorporation of loanforms in accordance with this system will result not only in rendering them harmonious with the native vocabulary but also capable of producing further derivatives. Evidence in support of this view can be derived from the fact that early instances of analogically arabicized borrowings, unlike cases of outright transfer, are now hardly distinguishable from native elements, and have given birth to many derivatives. For example the root q-n-n abstracted from *qa:nu:n* from Greek κανων (Eng. canon) has yielded the following forms:

qannan-	to legislate
muqannin	legislator, lawmaker
muqannan	determined, fixed; formed in accordance with the law
qa:nu:niyy	canonical, lawful, legal
qa:nu:niyyah	lawfulness, legality
taqni:n	legislation, lawmaking
qawa:ni:n	laws

Another case in point is l-j-m abstracted from *lija:m* from Per. liga:m "bridle", which underlies the following derivatives:

?aljam-	to bridle, rein in; to silence, restrain
?inlajam	be bridled; be restrained
lajamah/mulajjam	part of horse's cheek covered by bridle
malju:m/muljam	bridled, restrained
?ilja:m	bridling; silencing; curbing
?istaljam	to ask (someone) to bridle a horse
lajja:m	one who makes bridles
lujm/lujum/?aljimah	bridles

On the other hand, loanforms that have not been subjected to analogical arabicization have failed to produce further derivatives, let alone their incompatibility with the Arabic character. Let us illustrate this statement with one of the borrowings of modern times. The current word for "television" in Arabic is the loanform *tilivizyo:n* from French télévision, which is clearly out of line with the Arabic structure. The idea inherent in the verb "televise" has commonly been conveyed not through a verb, but a syntactic expression involving the use of the loanform itself. For example, a sentence like "the match was televised" would be rendered *nuqilati l-muba:ra:tu bit-tilivizyo:n* (lit. "the match was transmitted by television"). However, the root t-l-f-z has recently been abstracted from the same loanform, and new derivations based on it are gradually gaining currency. There are thus *talfaz-* "to televise", *talfazah* "television", *tilfa:z* "television set", etc.

The reader will have noticed that the new forms derived from abstracted roots are of typical Arabic structure. Apart from the fact that their roots are abstracted from non-native forms, there is nothing to set them apart from native formations. In view of this fact it is possible to argue that analogical arabicization gives Arabic an advantage over other languages where the process of loanword integration is not carried out along the same lines, or at least not with the same degree of systematization, and where loanwords are therefore often not so difficult to distinguish.(45)

(iii) Analogical arabicization represents the native speakers' intuitive knowledge of the mechanics of their language. The principles governing the process of word formation in Arabic and lexical patterning in particular cannot be seen as merely rules forced upon the language; rather they represent an important aspect of the speakers' linguistic intuition. Evidence for this can be found in the following facts:

(a) Despite the fact that borrowing has been largely a laissez-faire process, a large majority of the foreign vocabulary of the language achieves excellent harmony with its native patterns. This helps us draw the conclusion that the modificatory aspects of the borrowing process are underlain by the speakers' instinctive and unconscious compliance with the tendencies of their language. The idea embodied in this conclusion manifests itself clearly in the tendency on the part of native speakers to render the consonantal structure of loanforms to tri- or quadriconsonantal roots, a process whereby the given loanform, depending on the nature of its deviance from the native system, is either trimmed or built up, so that an appropriate form (or root) may be abstracted from it. This tendency is not confined only to the written form of the language; abundant evidence is also found in the spoken or colloquial variety, and it is from the latter that we shall take our examples, as borrowing in this case is an entirely natural process. The following cases have been attested in Iraqi Arabic, where foreign borrowings are reduced to conveniently shorter forms consisting of three or four consonants, which is usually the case in native vocabulary:

concrete	*kankari*
accelerator	*sikle:ta*
carburettor	*ka:bre:ta*
self-starter	*silf*
water join	*ja:yin*

There are also cases where new roots are abstracted from loanforms, thus giving birth to new derivatives which exhibit wide correspondence to SA morphological patterns. For example:

nervous n-r-f-z :
narfaz- "to make nervous" Fa8LaL-
tnarfaz- "to get nervous" *tFa8LaL- (cf. taFa8LaL-)
mitnarfiz "nervous" *mitFa8LiL (cf. mutaFa8LiL

cigarette j-g-r :
jiga:rah "cigarette" Fi8a:Lah
jaggar- "to smoke a cigarette" Fa88aL-
jaga:yir "cigarettes" *Fa8a:yiL (cf. Fa8a:?iL

goal g-w-l :
gawwal- "to score a goal" Fa88aL-

chance č-n-ṣ :
čannaṣ- "to have luck" Fa88aL-

On the other hand, an Arabic replica of a foreign model may have added to it an additional consonant which does not exist in the original form. This is especially the case where the loanform is biconsonantal. The addition of a third radical to such forms reflects the tendency of the speakers to respond to the mechanical demands characterizing the structure of their own language. It has been mentioned that the overwhelming majority of Arabic roots consist of three radicals, which accounts for this phenomenon. To illustrate the preceding statements, we refer to some of the examples given previously. For example, Per. čak "cheque" and kaz "coarse silk" are rendered *ṣakk* and *qazz*, respectively, where the final consonant is doubled in the Arabic replicas. Similarly Gr. μνα is arabicized as *mann* "weight of two pounds". The added sound in these cases is more clearly manifested in such morphological processes as pluralization. The latter for instance is pluralized as *?amna:n*.

A relevant process where the speakers' intuition may be said to have manifested itself is the following. It will have been understood from our account of Arabic syllabic structure that no Arabic word or syllable is initiated with a vowel (see 1.4.i). We have seen that in the case of loanforms where such a syllable may occur at the beginning of the word, a consonant sound is always used to precede the vowel, thus creating a permissible structure of the type CV or CVC. The consonant used in this case is either a glottal /?/ or /h/, or a pharyngeal /ᶜ/ or /ḥ/. Here are some cases:

Eng.	electron	*?ilikitro:n*	CV-
It.	accademia	*?aka:di:miyyah*	CV-
Gr.	οπιον "opium"	*?afyu:n*	CVC-
Per.	anda:m "stature"	*hinda:m*	CVC-
Gr.	εντυβον "endive"	*hindiba:?*	CVC-
Gr.	αχατης "carnelian"	*ᶜaqi:q*	CV-
Per.	anzaru:t "eye ointment"	*ᶜanzaru:t*	CVC-

Similar cases may also be cited from dialectal Arabic. Speakers of Iraqi Arabic, for instance, use the following forms:

hingala:nah	angle-iron
ᶜantarna:š	international
ᶜanti:k	antique

And in Ḥasāwi Arabic we find:

hinš	inch
ḥafi:z	office
ᶜa:yil	oil (see Smeaton, p.87)

(b) Instances of back formation (back derivation) constitute another arugment in support of the above view. For instance, the early loanform *kaᶜk* "pastry, biscuit" (from Per. ka:k)(46) has first been taken as a collective noun (referring to the group of items denoted by this word as one whole). A new singular noun (of individuality) has however later been back-derived from this form, namely *kaᶜkah* "an article of pastry". This is clearly done on the analogy of such native forms as *tamr* "dates" (coll.n.) and *tamrah* "a date"; *naml* "ants" (coll.n.) and *namlah* "an ant", etc. Another case is the Turkish loanform *taba:ši:r* "chalk". Due to its formal similarity to native plural forms of the pattern Fa8a:li:L, e.g. *jara:ṯi:m* "microbes", a singular noun *tabšu:rah* "a piece of chalk" has been back-derived from it, on the imperfect analogy of e.g. *jurṯu:mah* "a microbe" Fu8Lu:Lah.

(c) The predisposition of the native speakers of Arabic to bring loanforms into line with what is characteristic of their own vocabulary can further be evidenced by the fact that foreign compound constructions are frequently rendered into single units which are describable in terms of being root-based formations. For example, in the jargon of Iraqi car mechanics, "flywheel" becomes *fala:wi:n*, "dashboard" *dašbu:l*, "big end" *bigin*, etc.

(iv) Analogical arabicization may also be argued for on the basis that it constitutes a stabilizing factor in the process of loanform integration. The fact that a loanform conforms to one of the established patterns of the language helps reinforce its structural identity in the speaker's mind. An important point to be mentioned in this respect is that the morphological patterns of Arabic help cover a certain deficiency of its orthographic system. Arabic short vowels or diacritical marks (u.a.i) are more often than not unprovided in its writing system, the Qur?an being the only exception where these marks are always used to ensure correct reading of the Holy Book. As far as native texts are concerned, mistakes in reading are therefore inevitable unless one understands beforehand what one is going to read. Regarding the pronunciation of non-native words, the situation, which is even more difficult, may be illustrated as follows.

The arabicized form of Fr. télévision is written with an initial sequence of consonants which, due to the nature of Arabic syllabic structure, would admit a certain number of short vowels to be inserted amongst them, so that the end result may be an acceptable structure. The problem here is that the native reader or speaker has no clue as to what particular short vowel to insert and in what position, for the loanform in question does not correspond to a native pattern on which it may be modelled. This being the case, there now exist a variety of pronunciations of the same loanform. In addition to the one already mentioned on page 114, i.e. *tilivizyo:n*, there are: *talavizyo:n*, *talivizyo:n*, and *tilvizyo:n*. Notice that the sound /v/ of the original, although written /f/ (ف) in the arabicized version, almost always shows transfer, i.e. it is pronounced /v/ rather than /f/. The same may be said about the sound /o/, which is pronounced /o:/ although written /u:/(و).

On the other hand, the analogical form *tilfa:z* coined from the same loanform (p.114) is not likely to allow for similar variation since it corresponds to many similarly patterned native forms, e.g. *sirba:l* "shirt". Notice that in this case the non-native sound /v/ of the original form never shows transfer, which indicates that complete phonological assimilation goes hand in hand with full morphological integration.

4.4.2 Factors limiting the applicability of analogical arabicization

What has been said in 4.4.1 above may have given the impression that the process of analogical arabicization is capable of handling all types of loanforms and in all circumstances. Our intention, however, has not been to give such an impression. There are a number of linguistic as well as extralinguistic factors which we will now discuss.

4.4.2.1 Factors relating to the nature and controllability of the borrowing phenomenon

In view of the fact that loanforms find their way into the language through many and mostly uncontrollable channels, the occurrence of non-analogical loanforms is inevitable. The conscious efforts made by language academies, educational organizations and other concerned bodies, which are aimed at replacing these loanforms by native equivalents — or, when inevitable, rendering them consistent with the native system — have so far fallen short of realizing the desired objectives. The difficulty involved in checking and controlling the continuous influx of loanforms cannot, of course, be underestimated: especially in scientific and technical fields, the language is confronted with new importations almost daily. Attempts at furnishing native equivalents for these importations are far too slow to prevent them from penetrating the language. The speakers at large would not normally wait until they are told by academicians or language authorities what terms they should use, or in the case of loanforms, how they should pronounce them.

The fact that borrowings into Arabic come from more than one source constitutes another aspect of the difficulty referred to, especially if we take into account the fact that despite their recent attempts to co-ordinate their efforts, Arab countries are still far from reaching this goal. The Arab world is divided into two main sectors with regard to the foreign language used in teaching at universities and higher levels of education and in the use of bibliographic material. Thus, whereas French is predominant in North African countries, English has the lead in the rest of the Arab world. This state of affairs has resulted in various types of terminological discrepancy. Sometimes the same loanform is rendered differently in different Arab countries. A case in point is the French word "cadre", with the meaning "the higher staff (of an administration, a firm or a governmental agency)". In Iraq this word has been analogically arabicized as *al-ka:dir* with more or less the same meaning. Moroccans, on the other hand, have followed a different route to convey the same idea. They have used as a native substitute the word *?iṭa:r* (pl. *?uṭur*), which in fact is a translation of a different, though related, sense of the same loanform, namely "frame, framework", thus giving Ar. *?iṭa:r* a new aspect. We have also seen cases where two versions of the same loanform may exist in Arabic at the same time because it occurs both in English and French, from which it is borrowed (see p.78).

4.4.2.2 Factors relating to the type of loanform used and the field it relates to

The application of analogical arabicization does not seem always to be subject to the borrower's tendency or nature of linguistic orientation. A lot also depends on the type of meaning it is intended to convey and the particular domain of language it belongs to. Writers on arabicization often fail to realize a number of important distinctions involved in the use of words in ordinary or literary language, on the one hand, and in the specialized fields of science and technology on the other. Awareness of these distinctions is bound to help us provide answers to such questions as: Why are scientific borrowings relatively more resistant to formal as well as semantic changes than literary words? Why is borrowing more common in science than other domains of the language? What motivates the scientist when he shows a preference for foreign terms instead of the native material at his disposal. In order to answer such questions we need to know what distinguishes the language of science from ordinary language and, through that, what are the basic characteristics of scientific vocabulary. These issues represent the core of the following discussion.

4.4.3 The language of science and ordinary language

Let us think of a situation where two writers, a literary one and a scientist, are presented with the same proposition: a description of a common object, say a flower. A comparison of the language these two writers use in expressing their responses is bound to reveal certain aspects of difference in their attitudes. The description provided by the literary writer may reflect a purely artistic attitude where language may be employed in a most eloquent fashion to portray such aspects of the flower as its beauty, freshness, colour, fragrance, etc. His choice of words is largely influenced by such factors as his degree of literary sophistication, his preference for direct or figurative language, and even his state of mind at the moment. Subject to his instantaneous impression, he may find a particular word more expressive or more appropriate than another, although another literary writer might prefer a different expression. Nevertheless, his account will still be taken as a description of a flower. Furthermore, he does not expect his readers or listeners to attach to his words exactly the same meanings as he intends.

The scientist, on the other hand, is interested rather in the flower as part of a plant and looks at it from the point of view of a botanist. His main concern is to give a clear, accurate and precise description of it, where facts will not be questionable and are not capable of more than one interpretation. Unlike the literary writer, he expects his readers to receive his words with exactly the same meanings he attaches to them so that his message may be received without any possibility of misunderstanding or ambiguity.

We thus understand that a scientist uses his words as tools for exact and logical communication; no ambiguity or equivocation is allowed for. Monosignificance, clarity, precision and directness are essential requirements of his language. Whewell's definition of this language is well worth quoting in this connection (as quoted by Sheard, p.256):

> When our knowledge becomes perfectly exact and intellectual, we require a language which shall also be exact and intellectual; we shall exclude alike vagueness and fancy, imperfection and superfluity; in which each term shall convey a meaning steadily fixed and vigorously limited. Such is the language of science.

Another statement where the relation between words and ideas in scientific language is most clearly illustrated is that of A. Lavoisier, who, discussing the nomenclature of chemistry, has this to say (quoted by Adams, pp.209-10):

> We shall have three things to distinguish in every physical science: the series of facts that constitute the science; the ideas that call the facts to mind; and the words that express them. The words should give birth to the idea; the idea should depict the fact.... The perfecting of the nomenclature of chemistry ... consists in conveying the ideas and facts in their strict verity, without suppressing anything that they present, and above all without adding anything to them; it should be nothing less than a faithful mirror; for we can never too often repeat that it is not Nature nor the facts that Nature presents, but our own reasoning that deceives us.

From the point of view of ordinary language, on the other hand, while words with clearly defined meanings do exist, this is certainly not the case with the bulk of the vocabulary. Ordinary words often tend to be imprecise and have more than one meaning. One of the facts established in modern linguistic studies is that as a word extends its areas of application, it may gradually develop new or additional meanings which may be quite remote from the original one. Words may change their meaning as a result of the various types of change taking place in the community where these words are used (for illustrations in Arabic, see Anīs, 1976). The concept or idea with which a word is associated in the speaker's mind may thus vary with the particular time, need, circumstance, etc. calling for its use – this is what philologists call "change of associations". Sheard represents these semantic changes by "a series of concentric circles, the innermost enclosing the basic idea, and the outer circles becoming more unstable as they expand, needing more and more help from context if the meaning is to be grasped" (1954, p.256).

Now that we have formed an idea about the type of language associated with scientific expression, on the one hand, and that used for ordinary purposes, on the other,(47) let us specify and bring into focus those characteristics of scientific words which constitute the crux of the matter in our present discussion.

4.4.3.1 Characteristics of scientific vocabulary

Words used in the language of science have the following characteristics:

(i) Literary writers often complain that scientific words are "ugly" and discordant with the literary taste. Indeed scientists do not refrain from admitting this fact, but nevertheless they do not reject their words on this basis. That is because they do not concern themselves so much with the shape or beauty of the word as its efficiency for accurate and scientific communication. Arab writers who are in favour of avoiding the use of foreign borrowings, and those who advocate analogical arabicization when borrowing is inevitable, often base their views on the argument that borrowed scientific and technical terms lack the relative shortness, pleasantness and other aesthetic values characterizing Arabic native vocabulary. The fact that Arabic words, those formed through analogical derivation, generally tend to be short is not denied, but it remains to be seen whether it is always possible to modify scientific borrowings in accordance with the native system or native aesthetic values. On the other hand, the fact that a scientific term is characterized by unfamiliarity, foreignness or any other characteristic that sets it apart from ordinary words is not always looked upon with disfavour, at least not by scientists who use it. On the contrary, some take it to be an advantage. Referring to the situation in English, H. Bradley states (1920, pp.108-9):

> It is often a positive disadvantage that a scientific word should suggest too obtrusively its etymological meaning. A term which is taken from a foreign language, or formed out of foreign elements, can be rigidly confined to the meaning expressed in its definition; a term of native formation cannot be so easily divested of misleading popular associations. If, for example, the English founders of the science of geology had chosen to call it "earth-lore", everyone would have felt that the word ought to have a far wider meaning than that which was assigned to it. The Greek compound, which etymologically means just the same thing, has been without difficulty restricted to one only of the many possible applications of its literal sense.

To cite a relevant case from Arabic, an attempt has repeatedly been made by some purists to replace the ancient loanform *al-fi:zya:?* "physics" by the native word *aṭ-ṭabi:ᶜah* or *ᶜilmu ṭ-ṭabi:ᶜah* "science of nature", but the latter has been widely rejected on the grounds that it is too general, comprising all that exists in nature: animals, plants, minerals, etc. Similarly *al-faslajah* "physiology" is still predominant despite attempts at replacing it by *ᶜilmu waḍa:?ifi l-?aᶜḍa:?* "science of the functions of organs".

(ii) Unlike ordinary words, scientific terms do not normally change their meaning or acquire new associations in the course of time, for once they do so they lose their distinction as precise and unambiguous labels. This characteristic has led some philologists to describe science as the enemy of language. The implication is that the constancy characterizing scientific words runs counter to the general conception of language as an ever-developing and growing mechanism. Or, as Savory puts it (1953, p.52), science is seen "as though it were a kind of linguistic hormone that inhibited growth and development". However, mention must be made of the fact that, on passing into everyday language, purely scientific words may become more and more liable to semantic change as they begin to have new associations or (to use Sheard's metaphor) when they begin to expand their outer circles. But even then the scientist's main concern remains confined to the basic meaning of the word, its innermost circle. A case in point is the Arabic word *?inṣiha:r* "melting, fusion" (ṣ-h-r "to melt"), whose meaning is purely scientific. However, due to the potential metaphor suggested by the state of merger and absorption resulting from the process denoted by this word, it has now come to signify the social process whereby people, parties, sects, etc. of varying affiliations and affinities are called upon to unify against a common danger or for a common objective. A political slogan often raised these days is *al-?inṣiha:ru fi: bu:daqati n-niḍa:l* lit. "melting in the crucible of struggle (against the common enemy)". As far as the scientist is concerned, however, this figurative application of the word is absolutely irrelevant.

(iii) In view of the fact that scientific loanforms in present-day Arabic come mostly from English and other genetically related languages, it is essential that we acquaint ourselves with the characteristic type of formational processes involved in producing these words in the languages in question.

A cursory look at any English scientific book or dictionary would soon reveal the fact that the greatest majority of the words are coined out of formative elements taken from classical languages, Greek and/or Latin, which are welded together so that the resulting form is, so to speak, a manufactured or artificial word. Examples of such elements include aqua-, arch-, auto-, phono-, photo-, pyro-, -gram, -graph, -itis, -logy, -meter, -scope, -stat, -therapy, etc. Resort to this method has been due largely to the huge and rapid advance in science in the last hundred years or more, so much so that English has found its native resources not quite sufficient for its lexical needs. Scientists have thus found themselves compelled to create their own terms by drawing on the clasical languages, even if that meant the use in one and the same term of elements belonging to more than one language. Thus are obtained "haemoglobin" from Gr. haima "blood" + L. globulus "a small sphere"; and "micronucleus" from Gr. mikros "small" + L. nucleus "a nut", etc.

By creating such forms and by resorting to such methods, the scientist makes a clear manifestation of his primary concern with the scientific appropriateness of the words he uses rather than other considerations, such as shape, length, beauty, etc. which, from the point of view of the literary man, may be regarded as of greater importance.

(iv) Another characteristic of scientific terms is that they tend to have an international character. Many such terms are found nowadays in many of the languages of the word, sometimes regardless of whether or not they are genetically related. Here are some examples:

English	French	German	Arabic
transistor	transistor	transistor	*tranzistar*
magnetism	magnétisme	magnetismus	*magnati:siyyah*
electronic	électronique	elektronen	*?ilikitru:niyy*
dynamic	dynamique	dynamisch	*di:na:miyy*
oxidation	oxydation	oxydation	*?aksadah*

In addition to the factors mentioned above, the occurrence of scientific terms in many languages may be explained by the fact that scientists from different parts of the world are familiar with these terms and therefore find it easier and more practical to use them in their communication rather than native counterparts which may not be equally widely familiar.

4.4.4 The place of analogical arabicization in modern scientific terminology

It could be stated in general that scientific terms are less susceptible to the principle of analogical arabicization than ordinary words. This is particularly the case in chemical terminology. This seems to be largely due to the formational nature of these terms on the one hand, and the requirements of the principle in question on the other. We have seen that scientific terms are predominantly compound words, and that the elements constituting these terms are in most cases equally essential to its overall meaning, which rules out the possibility of omitting any of them (see 3.3-3.4; 4.3.3.1.iii). Analogical arabicization, on the other hand, is of necessity a process whereby the foreign model is subjected to various types of formal change, as has been amply illustrated in our previous discussions. Such changes would thus inevitably affect the semantic content of the scientific term and render it less capable of representing its original meaning accurately and precisely. That the constituents of the scientific term are equally indispensable may be seen from the fact that when such terms are replaced by native counterparts their constituents are fully reproduced. For example:

spectrograph	*mirsamatu ṭ-ṭayf*
thermostat	*muṯbitun ḥara:riyy*
geomorphology	*ᶜilmu šakli l-?arḍ* (48)

Consider also the following partially reproduced loanforms:

psychodynamic	*di:na:miyyun nafsiyy*
bichloride	*ṯa:ni: klo:ri:d*
spectrophotometer	*fo:to:mitrun ṭayfiyy* (48)

In view of the above facts, when fully borrowed into Arabic, scientific loanforms of the type just given show relatively faithful transfer, i.e. they are not analogically arabicized.

magnetometer	*maǵni:to:mitr*	
centimetre	*santimitr*	
microscope	*mikro:sko:b*	
fotografia (It.)	*fotu:ǵrafiya:*	"photography"
geophysical	*jiyofi:zi:qiyy*	
troposphere	*tro:po:sfi:r*	
pyrophosphoric	*bi:ro:fusfo:ri:k*	
hydrocarbon	*hi:dro:karbo:n*	
chloroform	*klo:ro:fo:rm*	
protoplasm	*pro:to:bla:zma:*	

We may well mention here the fact that as a by-product of the use of non-analogically arabicized loanforms, such as the ones just cited, there now exist in Standard (especially Scientific) Arabic features that are not characteristic of its system. Our preceding examples include a number of instances where consonant clusters occur at the beginning of Arabic renditions. There is also an increasingly noticeable tendency on the part of users of non-analogical scientific loanforms to retain non-native sounds which occur in the foreign models, e.g. /p/, which is revealed by some of the cases already mentioned, and /v/, as in *vi:ta:mi:n* "vitamin". The non-native vowels /o/ and /o:/ are also showing transfer, although some conscious speakers tend to replace them by the native /u/ and /u:/ respectively.

It should be mentioned however that there are certain cases where loan-roots have been abstracted from purely scientific words, thus making it possible to apply the principle of analogical formation. In such cases the loanforms involved are usually not compounds but secondary stems consisting of a base and a derivational affix. The consonantal elements constituting the base, or their nearest Arabic counterparts are used in coining the loan-root, which in turn is used to produce a derivative equivalent to the foreign model. Here are some examples from chemical terminology:

racemize	r-s-m	*ra:sam-*	Fa:8aL-
pasteurized	b-s-t-r	*mubastar*	muFa8LaL
sulfitation	s-l-f-t	*salfatah*	Fa8LaLah
ionization	?-y-n	*ta?ayyun*	taFa88uL
polymerism	b-l-m-r	*balmariyyah*	Fa8LaLiyyah
oxidant	?-k-s-d	*mu?aksid*	muFa8LiL
galvanizer	j-l-f-n	*mujalfin*	muFa8LiL

Notes to Chapter 4

1 al-Jawāliqiyy (1969) p.53. For a discussion of the origins of these words, see Jeffery (1938) pp.164-5, 266, 293, 206-7, and 58-60 respectively.

2 For a detailed discussion of these issues see, for instance, al-Suyūṭiyy (1967) vol.2, ch.38, and (1438 AH); also Jeffery (1938).

3 Bayt al-Ḥikmah was an academic institution which comprised a well-stocked library and an astronomical observatory and in which scientists from various parts of the world were well provided with all the facilities necessary for their life and studies.

4 For a detailed account of the efforts of early Muslim translators at rendering foreign books of science into Arabic, see: Ṭāha (1976) p.217 fn.2; Zaydān (1904) vol.3; O'Leary (1949). For another account of the influence of Arabic civilization on the world, the reader may refer to O'Leary (1954) and Goichon (1969) p.81, fn.2.

5 *handasah* is an analogically arabicized form of Pah. anda:za.

6 The new form *manjani:q* corresponds to the native pattern Fa8LaLi:L.

7 al-Jāḥiḍh (1961, vol.1, pp.141-2); the writer here cites a number of verses containing non-native words, which, as he puts it, are used by Arab poets to produce a pleasant effect (*lit-tamalluḥ*).
Ibn Qutaybah (1300 AH, pp.101-3, 176-8); he discusses the grammatical status of loanwords in Arabic and also quotes a number of verses where borrowed words are used.
Ibn Durayd (1344 AH, vol.3, pp.499-503; in his account he describes the way a number of foreign borrowings have been so fully naturalized that they have become part of the language. He also substantiates his argument with verses containing these borrowings.
Ibn Jinniyy (1913, vol.1, pp.362f). See also 1954, p.242, fn.
al-Jawhariyy (1375 AH, p.204 fn.).
al-Thaᶜālibiyy (1938, pp.314-17).
Ibn Sīdah, "al-Mukhaṣṣaṣ", vol.4, part 14, pp.39-44.

8 al-Khafājiyy (d.1659) (1325 AH, p.3); see also al-Karmiyy (1921, p.131).

9 al-Ḥarīriyy (1299 AH, p.80): *?iḏa: ᶜurriba l-?ismu l-?aᶜjamiyy, rudda ?ila: ma: yustaᶜmalu min naḏa:?irihi fi: luġatihim waznan wa ṣi:ġatan.* Echoing the same view, al-Afghāniyy (d.1898) is reported to have said "if we want to use a foreign word, all we have to do is dress it in a *mašlaḥ* (cloak) and *ᶜiqa:l* (headband), thus rendering it Arabic". See al-Maghribiyy (1947, p.2).

10 al-Ḥarīriyy, p.39; this word derives from Per. havang, meaning "mortar (for pounding)".

11 Meaning: very timorous; also, he who distinguishes truth from falsehood (epithet of the Second Caliph ᶜUmar).

12 al-Ḥarīriyy, p.61; originally Per. dastu:r "constitution".

13 Meaning: buffoon, jester.

14 al-Ḥarīriyy, p.29; originally Per. sard-a:b "(place for) cold water".

15 Meaning: garment.

16 al-Ḥarīriyy, p.62; this word is reported by Addi Shirr (viz. fn.3 p.256; p.20) as deriving from Pah. partalah, meaning "a long piece of iron"; however, cf. al-Jawāliqiyy, fn.3 p.116.

17 Meaning: brave, valiant.

18 al-Ḥarīriyy, p.80; originally Per. šatrang, "chess".

19 Meaning: stout camel.

20 Referring to the interrelation between language and other aspects of culture, Hoijer points out that this interrelation is "so close that no part of the culture of a particular group can properly be studied without reference to the linguistic symbol in use". He also adds that "linguistic change tends to slow down where the culture of a people is relatively static or slow to change, and that, when a group undergoes rapid changes in its non-linguistic culture, linguistic change may similarly increase in tempo". See Hoijer, 1964, pp.456-8.

21 See Chejne (1969, p.66) and Sapir (1921, p.194). The latter describes Arabic as one of five languages that have had an overwhelming significance as carriers of culture, the four others being classical Chinese, Sanskrit, Greek and Latin.

22 Different writers may vary in arranging the various linguistic domains in the order in which they are subject to interlingual influence, but it is observed that it is always the vocabulary that comes first. See Weinreich, 1968, p.67.

23 al-Jawāliqiyy, p.60. Earlier statements to the same effect were made by Ibn Durayd (1344 AH, vol.1, p.9) and Sībawayh (1361 AH, vol.2, p.460).

24 See Greenberg (1950, p.163). This study is in fact mainly based on Arabic because, as the writer puts it, "of the abundance of lexicographical information and the relative archaism of its phonological structure".

25 Ibn Jinniyy (1954); this book, which is often referred to by modern writers under the title Sirr al-Ṣināᶜah, is the topic of a thesis by Muhammed Hasan Bakalla (University of London, SOAS, 1970).

26 This form was later pronounced with /s/ rather than /z/, i.e. *muhandis*, which is more in line with the Arabic system.

27 For an interesting discussion on these consonants see al-ᶜUbaydiyy (1979, II, pp.291-327).

28 Ibn Jinniyy, 1954, 1, p.74.

29 By way of giving this loanform an Arabic shape, al-Karmaliyy has assigned it the pattern Fa[c]alu:n, i.e. *talafu:n*, on the analogy of *ḥalazu:n* "spiral". See his "Nushū? al-Lughah", p.97.

30 See al-Fayrūzabādiyy, 1301 AH, vol.1, p.128; cf. al-Bustāniyy, 1870, vol.2, p.1928.

31 The loanform *film* is pronounced with a light rather than a dark /l/, unlike the case in the foreign model.

32 According to Bāqir, this word, which entered Arabic via Greek, is ultimately of Sumerian origin, and around 2500 BC it was used as a title meaning "the king of the region (Sumer)" (lu:ka:l kala:ma:). See Bāqir's article concerning this word in the Bulletin of the Faculty of the Arts, University of Baghdad, vol.xxiv, January 1979, pp.509-56.

33 al-Suyūṭiyy, "al-Muzhir", 1, p.274.

34 See Shirr, 1908, p.20.

35 al-Suyūṭiyy, "al-Muzhir", 1, p.274.

36 al-Yasū[c]iyy, 1960, p.241.

37 The following examples are also of Persian origin. It should be noted that our use of the designation "Per(sian)" is not intended to imply a specific stage of the Persian language during which words so designated have been borrowed by Arabic. A certain borrowing may thus trace its origin back to Pahlavi, i.e. Middle Persian, rather than later stages. However the specific designation "Pah(lavi)" will also be used when this is found necessary.

38 See Steingass, 1892, p.790, and al-Yasū[c]iyy, 1960, p.237.

39 Steingass, p.402; cf. al-Yasū[c]iyy, p.372.

40 al-Yasū[c]iyy, p.224; see also Ma[c]lūf, 1966, p.105.

41 Steingass, p.637; cf. Shirr, p.76.

42 Steingass, pp.968, 1027; see also al-Jawāliqiyy, p.321; Ma[c]lūf, p.626; al-Maghribiyy, p.32.

43 See Blochet, 1896, p.271; see also Ma[c]luf, p.362.

44 See Browne, 1921, p.34.

45 In her study of linguistic borrowing in Norwegian, Aasta Stene has observed that "foreign words tenaciously keep characteristic foreign traits. Even if they superficially look native, some formal characteristic ... betrays their foreign status." (1945, p.210).

46 Here we have another case where a biconsonantal loanform is rendered triconsonantal.

47 For an interesting and more detailed account, see Savory, 1953.

48 Many such examples are provided in Chapter 3, esp. 3.4.

5 The questionnaire

Introduction

On 25 February 1980 I travelled on a fieldwork visit to Rabat, Morocco, where I stayed for two weeks, during which I collected data and visited the Permanent Bureau of Co-ordination of Arabicization (PBA) and the Institute of Researches and Studies of Arabicization, both of which are almost exclusively concerned with the development of the Arabic language. The visit to these two organizations offered a fruitful opportunity to meet with and consult a number of the most prominent scholars concerned with the subject of this book. I was also particularly lucky that my visit to Rabat coincided with the convening of a linguistic colloquium (4-7 March 1980) concerned with writing Arabic books for speakers of other languages. The PBA very kindly invited me to attend this colloquium in which a large number of leading experts in Arabic (non-Arabs as well as Arabs) took part. It was thus a rare opportunity to consult and exchange views with as many of those scholars as possible.

Prior to this visit to Rabat, I had paid a similar visit to Iraq, my home country, from 4 January to 25 February 1980, during which I conducted a questionnaire at the University of Baghdad. This chapter is devoted to a discussion of the reasons and objectives of this questionnaire, as well as the results and findings it has helped to achieve.

5.1 Reasons and objectives

In the preceding chapters we have discussed a number of methods of lexical creation and neologization. We have seen that the results (i.e. the lexical creations) yielded by these methods vary, sometimes greatly, with regard to more than one fact: their correspondence to the morphological system characteristic of Arabic; their practicability in actual linguistic practice; their correspondence with the current tendencies of present-day speakers, etc. Besides, as has been observed above, some of the efforts carried out in this field are better described in terms of being theoretical attempts by lexicographers and scientists concerned with the question of overcoming the terminological difficulties experienced when using Arabic in scientific and technological disciplines. The fact that many such attempts have not yet been put to the test of actual linguistic practice accounts for the prolonged and heated controversy around their practical values. An important factor disregarded by such attempts concerns the role of the speakers, their linguistic intuitions, and the types of coinages they would be likely to adopt. Such attempts often fail to recognize that a decisive factor in the currency of a linguistic form is its acceptance by the speakers.

The reader will have noticed that we have tried to give our own account of the various issues arising from our analysis of these attempts. Our main concern has been to view things in their correct perspective, taking into consideration every aspect of these phenomena and basing our judgements on objective and realistic considerations as far as possible. However, since some of the proposed methods and the formations they yield are no more than suggestions hardly confronted outside the context in which they are put forward (or, in other instances, their limited occurrence has given rise to much controversy about their practicality), we have found the questionnaire useful to fulfil the following objectives:

(a) To test the acceptability of the various types of lexical formations included in our study, especially those around which most of the controversy has revolved. This also helps us to derive certain conclusions concerning the speakers' intuitive knowledge of what may or may not be concordant with the developmental aspects of their language, as well as their linguistic taste (although this is to some extent a relative matter).

(b) To assess the speakers' tolerance for linguistic innovation, which reflects the extent and type of linguistic adaptation occasioned by the need to cope with modern requirements for scientific expression.

(c) To determine the type of factors accounting for the acceptance on the part of the speakers of certain types of word-formational methods at the expense of others.

(d) To see how far the results of the questionnaire correspond to the views and arguments previously put forward.

5.2 About the questionnaire

The people responding to this questionnaire were university students at Baghdad University, reading for the degrees of BA and BSc in four different departments. Table 3 shows their colleges, years of study, departments and numbers.

Table 3. Questionnaire respondents

	College	Year of Study	Department	Number
1	Arts	Fourth	Arabic	27
2	Arts	Third	English	26
3	Education	Third	Chemistry	25
4	Sciences	Fourth	Physics	24
			Total:	102

As can be seen from the table, the students fall into two main categories. Groups 1 and 2 belong to the category of human sciences while groups 3 and 4 belong to that of natural sciences. The inclusion of both categories was thought to be necessary since one of our main objectives is to study the type and extent of adaptation involved in the use of Arabic as a scientific medium as compared with its use as a literary or ordinary language. It is also hoped that the results will help us know how speakers of Arabic with varying fields of interest react to certain developmental or innovational aspects of their language.

Groups 1 and 2 involve students specializing in Arabic and English respectively. The purpose behind this is to see how far a good knowledge of Arabic, or, on the other hand, acquaintance with and interest in a foreign language, affects the speakers' attitudes and judgements regarding questions such as adherence to native-based neologisms or methods of lexical enrichment and particularly arabicization and the incorporation of non-native elements.

The questionnaire consisted of seven pages. The first page contains an introduction aimed at orientating the respondents with the problems, reasons and objectives underlying the questionnaire, as well as the task they are requested to perform. A full translation of the contents of this first page reads as follows:

A Questionnaire
about Samples of Scientific, Technical
and Cultural Terminology in the
Arabic Language

> The Arab speaker is well aware that the Arabic language in the present time is facing great challenges with regard to the vocabulary of science, technology and modern civilization. Opinions have differed both on the theoretical as well as practical levels concerning the capability of Arabic to face these challenges and its resources and possibilities in this respect.
>
> Such being the case, we find that attempts made at reviving the language have been varied and inconsistent. Among the reasons accounting for this situation is that many of those who have made these attempts have failed to pay due attention to a number of decisive factors involved in this matter. Many of these attempts are still in the form of suggestions not attested by real occurrence in actual linguistic practice, which accounts for the continued controversy around their significance and efficacy.
>
> The following pages contain samples of terms yielded by such attempts, which the speakers are kindly requested to examine, so that we may be able through consulting their opinions to determine what may be consistent with their linguistic intuition and their linguistic needs for the expression of scientific and technological concepts and ideas which are brought about by human civilization in our

present age, and which is bound to render Arabic a flexible tool in the service of scientific thought. In order that the meaning and linguistic function of the given term be clear, we have introduced it into an appropriate sentence or phrase; we have also supplied the alternative expression available to Arabic, especially when the suggested term constitutes a new or uncommon phenomenon in Arabic.

What the speaker is kindly requested to do is to point out to what extent he accepts or approves each of the models included in the following samples and as is required at the foot of each sample or opposite each instance. Notice that each sample represents an independent phenomenon.

We are undertaking this questionnaire due to our full conviction of the great significance of the role of the speaker and the nature of his needs in the growth and development that takes place in his language. We have great confidence that our questionnaire will be received with full attention and co-operation from all participants, to whom we would like to express in advance our deep gratitude for the time and effort they offer us.

This questionnaire is part of a PhD research carried out by A.S.M. Ali, at the School of Oriental and African Studies, University of London

Pages 2 through 7 contain the material of the questionnaire, which revolves around five different terminological issues, each represented by a sample of terms or coinages based on a certain principle of word formation. At the beginning of each sample there is a concise statement of the method and views underlying the coining of the terms represented by that particular sample. The following is a brief account of the different samples, with a single example taken from each by way of illustration.

Sample 1

This sample contains ten foreign terms belonging to a variety of fields, each of which has as its counterpart 2-4 renditions based on different methods of lexical creation. An example:

Foreign term	Arabic counterpart	Its nature and method of formation
Hydrolyse	a) *ḥallala bil-ma:?*	Arabic verbal construction consisting of the verb *ḥallala* + the noun *ma:?*
	b) *?ama:h-*	Arabic verb derived from the root m-w-h
	c) *halma?-*	Arabic verb coined (by Naḥt) from *ḥallal-* + *ma:?*
	d) *hadral-*	Analogically arabicized verbal form derived from the foreign term by abstracting the root h-d-r-l from it.

Although there is no space for it in the preceding example, the quesionnaire contained an additional column in which the respondents could indicate their degree of preference for each of the counterparts provided for the foreign term. This was done, as explained in a note at the top, by inserting number 1 between two brackets opposite to the equivalent considered most acceptable, then number 2 for the one ranking second in order of preference, and so on.

Sample 2

Sample 2 deals exclusively with the phenomenon discussed in Chapter 4, analogical arabicization, whereby a lexical root is abstracted from a foreign term and then used to yield one or more derivatives in accordance with the characteristic system of Arabic morphology. The instances included in this sample (five in number) are each represented by a single derivative which is given in an appropriate context illustrating its meaning and application. It is then replaced, in the same context, by a periphrastic expression employing the original loanform from which the above-mentioned root was abstracted. Both renditions are underlined. Here is an example:

Foreign term	(A) Analogical formations from borrowed roots, in context	(B) Alternative constructions involving the original loanform
televization Fr.télévision	*lam yakun bil-?imka:ni* <u>*talfazatu*</u> *l-muba:ra:h*	*lam yakun bil-?imka:ni* <u>*naqlu*</u> *l-muba:ra:ti* <u>*bit-tilivizyo:n*</u>

Sample 2 differs from Sample 1 in that in the former the five instances given represent one phenomenon and the respondents are therefore requested to treat them as a whole by answering the following question which is given at the end of the sample: (This procedure also applies in the case of Samples 3, 4 and 5.)

> To what extent do you accept the underlined lexical formations in (A)? Please put the mark (✓) against the appropriate answer:
>
> (a) Acceptable ( )
> (b) Acceptable to some extent ( )
> (c) Unacceptable ( )

Sample 3

The lexical formations included in Sample 3 belong to the type of constructions discussed in 3.4 above, and referred to therein by the term *al-murakkabu l-mazjiyy*. Briefly, this is a method whereby a lexical unit is created out of the joining together of two words both capable of independent existence elsewhere other than in the resulting combination. The respondents are requested to follow the same procedure as in the previous case. Here is an instance representative of the cases included in this sample:

Foreign term	Arabic compound term in context	Alternative expression
nasopharyngeal	*al-?iltiha:ba:tu l-?anfi:bulᶜu:miyyah ...*	*... al-mutaᶜalliqatu bi-minṭaqati l-?anfi wa l-bulᶜu:m ...*

Sample 4

This sample represents the type of constructions treated in 3.3.4 under the heading "Full (Opaque) N-constructions", where the reader will find a detailed account of the principles underlying this type of word formation. Here is an example of the cases included:

Foreign term	New Naht-construction in context	Equivalent Arabic construction
nephrostomy	*?inna l-fatkalata ᶜamaliyyatun jira:ḥiyyatun xaṭi:rah*	*?inna fatḥa l-kulyati ...*

It should be mentioned that in this case the choice was given only between two options: (a) acceptable or (b) unacceptable (i.e. the option "acceptable to a certain extent" is not included). This is because in our opinion this case does not allow for such an option.

Sample 5

Sample 5 is concerned with another type of N-construction, designated in Chapter 3 by the term "Partial (semitransparent) N-constructions", and more specifically that subcategory where the abbreviated constituent of the construction is a "closed-class item". This sample may be represented by the following example:

Foreign term	Partial N-construction in context	Equivalent expression
prehistoric	*yadrusu l-ᶜulama:?u ?a:ṯa:ra l-?insa:ni l-qabta?ri:xiyy*	*... ?insa:ni ma: qabla t-ta?ri:x*

5.3 Results and findings

The following is an account of the facts and findings derived from our analysis of the results received in response to our questionnaire. These results are presented in statistical tables for convenient reference.

Findings relating to Sample 1

A cursory look at the results shown in Table 4 (overleaf) will reveal what may be described as a general fact, namely that the acceptance of any term, whether native or non-native, relatively short or long, and whatever its method of formation, is subject to a variety of factors with varying degrees of decisiveness, and can hardly be said to be attributable exclusively to any one of the

Table 4. Questionnaire results (sample 1)

Degree of preference		Percentages of results 1				2				3				4			
Department		Ar	En	Ch	Ph	Ar	En	Ch	Ph	Ar	En	Ch	Ph	Ar	En	Ch	Ph
1 Hydrolyse	a) *ḥallala bil-ma:?*	74	73	88	71	15	19	8	25	11	8	0	4	0	0	4	0
	b) *?ama:h-*	18	12	8	13	52	35	60	21	26	30	28	29	4	23	4	37
	c) *ḥalma?-*	4	0	0	12	26	15	32	21	48	50	8	38	22	35	60	29
	d) *ḥadral-*	4	15	4	4	7	31	0	33	15	12	64	29	74	42	32	34
2 Helicopter	a) *al-helikoptar*	11	50	48	42	11	31	40	21	78	19	12	37				
	b) *al-ḥawwa:mah*	41	15	4	12	44	19	8	25	15	66	88	63				
	c) *aṭ-ṭa:?iratu l-ᶜamu:diyyah*	48	35	48	46	45	50	52	54	7	15	0	0				
3 Haemoglobinometer	a) *miqya:su l-himo:ġlo:bi:n*	33	69	96	79	15	8	0	13	52	23	4	8				
	b) *miḥma:r*	55	27	4	17	41	42	72	41	4	31	24	42				
	c) *miqya:su l-yaḥmu:r*	11	4	0	4	45	50	28	46	44	46	72	50				
4 Secretaire	a) *as-sekrete:r*	18	58	60	50	41	31	32	25	41	11	8	25				
	b) *an-na:mu:s*	19	0	4	8	22	15	8	17	59	85	88	75				
	c) *?ami:nu s-sirr*	63	42	36	42	37	54	60	58	0	4	4	0				

5	Barometer	a)	*miqya:su ḍ-ḍuġṭi j-jawwiyy*	44	39	68	50	41	42	23	37	15	19	8	13
		b)	*al-miḍġaṭ*	41	23	8	17	33	35	24	25	26	42	68	58
		c)	*al-ba:ro:mitr*	15	38	24	33	26	23	52	38	59	39	24	29
6	Electron	a)	*al-?ilikitro:n*	44	81	88	83	56	19	12	17				
		b)	*al-kuhayrib*	56	19	12	17	44	81	88	83				
7	Camera	a)	*al-miṣwarah, aṣ-ṣawwa:rah*	37	15	4	8	33	39	16	25	30	46	80	67
		b)	*al-ka:mirah, al-kamarah*	7	42	28	34	26	27	52	37	67	31	20	29
		c)	*?a:latu t-taṣwi:r*	56	43	68	58	41	34	32	38	3	23	0	4
8	Nitrogen	a)	*an-nitro:ji:n*	52	54	92	100	48	46	8	0				
		b)	*al-muxṣib*	48	46	8	0	52	54	92	100				
9	Radio	a)	*al-midya:ᶜ*	96	65	68	54	4	35	32	46				
		b)	*ar-ra:dyo:*	4	35	32	46	96	65	68	54				
10	Telephone	a)	*al-ha:tif*	89	69	76	92	11	27	20	8	0	4	4	0
		b)	*al-?irzi:z*	0	0	4	0	15	12	4	0	85	88	92	100
		c)	*at-tilifo:n*	11	31	20	8	74	61	76	92	15	8	4	0

characteristics just named. Evidence for this may be derived from the number of conclusions revealed by our analysis of the questionnaire results shown in Table 4.

One of the conclusions derivable from Table 4 is that subjects belonging to group 1, i.e. those specializing in Arabic, show in almost every case a greater tendency to favour native-based terms or coinages (referring to single-root lexical units only). Compare for instance the first options (column 1) in the cases: 1(b)-6(b), 7(a), 8(b) and 9(a)-10(a). This situation may be accounted for in terms of the fact that these respondents, being students of Arabic, are equipped with a more sophisticated knowledge of its morphological capability and lexical wealth than students of other disciplines, and are therefore more keen to promote attempts aimed at consolidating the role of native resources in meeting the demands for new vocabulary. As a matter of fact, it is generally observed that specialists in Arabic and native scholars educated in it are most often characterized by strong linguistic loyalty and a feeling that it is incumbent upon them to preserve the purity of the language at any cost. (This need not imply, however, that others are not loyal to the language or are totally unconcerned about its status; rather the matter is viewed in relative terms.) This may be substantiated in the present connection by the fact that, prior to our questionnaire, some of the coinages introduced in it were hardly known to the students. Yet students of Arabic exhibited a fairly high degree of preference for them when compared with students of other subjects. For example, the current Arabic counterparts for "haemoglobinometer", "electron" and "nitrogen" are almost exclusively the renditions *miqya:su l-hi:mo:ġlo:bi:n*, *al-ʔilikitro:n* and *an-nitro:ji:n*. However, the alternative newly introduced coinages: *miḥma:r*, *al-kuhayrib* and *al-muxṣib* were chosen as first preferences by 55%, 56% and 48% respectively of the respondents specializing in Arabic. (Similar cases are represented by 1(b), 2(b), 4(b), 5(b), 7(a) and 9(a).) Other respondents, however, showed a comparatively lesser degree of preference for these coinages (see Table 4, column 1).

At first sight case number 10(b) might be taken to represent an example contrary to the phenomenon just observed. For the term *al-ʔirzi:z*, which was proposed as an equivalent of Eng./Fr. "telephone", is also a native-based coinage deriving from an indigenous root r-z-z "to sound (especially of something distant)". Nevertheless, none of the students of Arabic opted for it as a first choice, only 15% accepted it as their second preference, and 85% took it as a last choice. An explanation for this may be that alongside *ʔirzi:z* there was another native alternative, *al-ha:tif* h-t-f "to call", which in earlier Sufism meant "an invisible caller or voice". Unlike the former, this has in recent years met with considerable success and is rapidly gaining ground. Evidence for this can be found in the fact that a fairly high percentage favoured *al-ha:tif* as their first preference (see 10(a)).

One reason why *al-ha:tif* has prevailed over *al-ʔirzi:z* seems to be that the root to which the former belongs (and other derivatives from it) is far more commonly employed than is the case with the latter. The overwhelming majority of present-day Arabic speakers, indeed, could have only a vague idea, if any, of what the coinage *?irzi:z* would be likely to stand for. The speaker is thus confronted with a word he is almost entirely unaware of. The same also applies in the case of *an-na:mu:s* "secretary", as opposed to *?ami:nu s-sirr* (4(b) and (c)).

As far as the preference for loanforms is concerned, respondents studying Arabic showed the lowest percentage, when compared with the other groups (see 1(d), 2(a), 4(a), 5(c), 6(a), 7(b), 8(a), 9(c), 10(c)). Reasons for this have been suggested above. Other respondents were relatively less reluctant to accept these loanforms. As can be seen in Table 4, Arabic has not been incapable of furnishing its own equivalents for these borrowings, so we must allow for other factors to account for their incorporation or preference at the expense of what is native. One reason why certain native equivalents have, at least so far, failed to compete with foreign borrowings seems to be that they were coined or introduced long after the latter had established themselves in the language and become familiar to the speakers. This is particularly true of the language of science, where a number of other factors make the replacement of such loans difficult (see 4.4.3-4.4.3.1). For instance, in physical and chemical terminology Arabic has always adopted the loanforms *?ilikitro:n* "electron" and *nitro:ji:n* "nitrogen". In classrooms, textbooks, journals and other forms of scientific communication, other alternatives have hardly existed. It should not be surprising then that a significantly high percentage of science students opted for them as first preferences; other subjects also exhibited a considerable degree of preference for such cases:

Loanform	Chemistry	Physics	Arabic	English
?ilikitro:n	88%	83%	44%	81%
nitro:ji:n	92%	100%	52%	54%

The same observation may be said to be true of cases where the loanform constitutes part of a hybrid more-than-one-word lexical unit. A case in point is the term *miqya:su l-hi:mo:ġlo:bi:n* "haemoglobinometer" (3(a)), where *hi:mo:ġlo:bi:n* is an arabicized form of the original "haemoglobin", while *miqya:s* is a native word meaning "measure". The native word *al-yaḥmu:r* "red", which has been suggested by some scholars as a counterpart for the former has hardly ever been actually used in the sense of "haemoglobin". This helps to account for the high tendency to use the hybrid construction mentioned above in preference to the fully native *miqya:su l-yaḥmu:r* (compare the statistics in 3(a) and (c)).

The conclusion to be derived is that the familiarity of a lexical unit is an important factor in its acceptance by the speakers, and that whether it is native or non-native cannot be said to constitute the only factor at work.

On the other hand, we have observed that when both the native and non-native forms are equally unfamiliar to the speaker, the general tendency is rather to adopt the native one at the expense of the non-native. For instance, the current term for "hydrolyse" is the verbal construction *ḥallala bil-ma:?*. The native *?ama:h-*

"(basically) to add water to" m-w-h, and the analogical loanform *hadral-* h-d-r-l "hydrolyse", are recent alternatives put forward by some modern lexicographers as relatively more economical forms, but neither of these two forms has occurred in actual linguistic practice. However, when presented in our questionnaire as possible alternatives (alongside *ḥallala bil-ma:?* and another Naḥt-fashioned one, *ḥalmaʔ-*), *ʔama:h-* was in general regarded as more acceptable than *hadral-*. The former was chosen by 13%, the latter by 7%; and while the latter was relegated to the category of last choice by 46%, the former was put last by only 17% (see 1(b) and (d)).

Our questionnaire also contains cases where Arabic can be said to employ two terms to designate the same object, one native and the other a loanform, and where both terms can be said to enjoy a certain degree of currency. A situation of competition is thus created between the two terms. It should be mentioned that in such cases the object denoted is usually a product of modern civilization. When cultural objects are borrowed they are usually accompanied by the names assigned to them in their respective cultures. These borrowed names may later have to compete with indigenous ones derived or neologized as substitutes for them. Instances are represented in our questionnaire by the designations currently used in Arabic for the artefacts known in English and several other European languages by the names "radio" and "telephone". For the former there now exist in Arabic the native coinage *al-miḏya:ᶜ* ḏ-y-ᶜ "to announce" and the loanform *ar-ra:dyo:*; for the latter there are the native *al-ha:tif* (see p.136) and the non-native *at-tilifo:n*.

As far as their occurrence as names of the above objects is concerned, the coinages *miḏya:ᶜ* and *ha:tif* are relatively more recent than their respective borrowed counterparts. However, due to their frequent recurrence in the mass media and in the works of some writers, these native names are becoming more and more popular. Another factor adding to their popularity may be that they are derivatives of widely employed roots in current Arabic and hence fairly familiar to the speakers. The loanforms *ra:dyo:* and *tilifo:n*, on the other hand, also still enjoy a more or less equal degree of popularity in both Standard and Colloquial Arabic, though more commonly in the latter.

The respondents to our questionnaire exhibited a relatively greater tendency towards native coinages rather than their borrowed counterparts; 72% and 81% opted for *al-miḏya:ᶜ* and *al-ha:tif* respectively as first preferences (see 9(a-b) and 10(a,c)). It may seem possible to generalize on the basis of this finding that, given the circumstances mentioned above, native equivalents such as these may be said to be more likely in the long run to predominate over their borrowed counterparts.

Another interesting fact revealed by our investigation which is of some relevance is that an old native word which is no longer operative, or a coinage based on an ancient root of no or very limited currency in present-day Arabic, will inevitably confront great difficulty in competing against a borrowed equivalent which, through common use and over a certain period, has succeeded in achieving some measure of familiarity and acceptance. A case in point is the aforementioned loanform *tilifo:n*, which was received far more happily than the native coinage *al-?irzi:z* (see p.136). The following table illustrates the percentages who opted for the foregoing alternatives as first, second or third preferences:

	1	2	3
tilifo:n	18%	75%	7%
?irzi:z	1%	8%	91%

It was suggested earlier that brevity is one of the characteristics for which a given term may be preferred. It has to be added, however, that when competing with another characteristic, namely familiarity, brevity seems to give precedence to familiarity. A relatively longer term with a high degree of familiarity or publicity seems to have a greater chance of predominance than a relatively shorter but less familiar one. Compare for instance cases 5(a-c); the multi-word unit *miqya:su ḍ-ḍuġṭi j-jawwiyy* "barometer", despite its relative length, was in general shown as preferable to the shorter native coinage *al-midgat* as well as the corresponding loanform *al-ba:ro:mitr*. The same also holds true in the case of *ḥallala bil-ma:?* "hydrolyse", as opposed to the natives *?ama:h-* and *ḥalma?-*, as well as the foreign-based *hadral-* (see 1(a-d); 3(a-c)).

Findings relating to samples 2-5

As pointed out earlier, samples 2-5 deal with four groups (each consisting of five instances) of lexical units based on different methods of word formation. An account of the nature of these formations is given on pages 132-3 above, and a more detailed discussion of them has been given earlier. We shall now try to give an analytical account of the results obtained with regard to these samples. The following table contains the percentages of the various responses of all the four groups of respondents.

Table 5. Questionnaire results (samples 2-5)

	Acceptable				Acceptable to a certain extent				Unacceptable			
Sample:	2	3	4	5	2	3	4	5	2	3	4	5
Arabic	55%	33%	22%	33%	41%	56%	-	41%	4%	11%	78%	26%
English	46	31	19	34	42	46	-	58	12	23	81	8
Chemistry	24	12	8	28	68	52	-	56	8	36	92	16
Physics	29	21	4	38	46	29	-	8	25	50	96	54

We may as well consider the attitudes of the respondents towards each of the types of morophological processes represented by the different samples. It is evident from Table 6 overleaf that sample 2 enjoys the highest degree of acceptance, whereas sample 4 lies at the other extreme. Samples 5 and 3 respectively rank second and third. The reader will notice that these results correspond fairly closely with our views expressed previously concerning the relevance and practicability of the principles underlying the lexical formations in these samples.

Table 6. Questionnaire results concerning the general acceptability of morphological processes (samples 2-5)

Sample/process	Acceptable	Acceptable to a certain extent	Unacceptable
2	39%	49%	12%
3	25%	46%	29%
4	14%	-	86%
5	33%	41%	26%

The relative acceptance of sample 2 coinages may be accounted for by the fact that as far as their morphological structure is concerned these coinages are absolutely harmonious with the native vocabulary, despite the fact that their roots are abstracted from non-native elements. From an aesthetic point of view they are thus as acceptable to the native ear as any indigenous words. This statement is supported by the fact that only 4% of the students specializing in Arabic did not favour this method, while 55% of them expressed their approval of it (see Table 5). From the point of view of its practicality, this method may be seen as advantageous in that it reduces to economical morphological constructs syntactic constructions of relatively greater length. Compare, for instance, *at-talfazah* with *an-naqlu bit-tilivizyo:n*, or *yukalwir* "chlorinate" with *yuᶜa:lij (yuᶜa:mil) bil-klo:r* "treat (cause to combine) with chlorine". The fact that neologisms based on abstracted (foreign) roots have existed in Arabic and can be traced back to very early stages of its history (see Chapter 4) may have been another encouraging factor contributing to the favourable opinion generally held towards coinages of this sort and the principles underlying their formation. Further arguments in favour of this procedure may be derived from our detailed study of the subject in 4.4.1 above.

Next on the scale of acceptance is sample 5. Constructions of the type *qabtaʔri:xiyy* "prehistoric" involve the initial affixation to a full word of a prefixal element (in the present case *qab-* < *qabla* "before") segmented from an otherwise independent preposition which is a full word in its own right. Similar cases contained in the sample include *faw-* < *fawqa* "above, super-", *taḥ-* < *taḥta* "under, sub;", and *bay-* < *bayna* "between, inter-". For additional examples, see 3.3.4 above and appendix.

It should be mentioned that so far Arabic has not witnessed the occurrence of these prepositions in their abbreviated forms. Words like "supersonic", "subcutaneous", "intercontinental", would be rendered *fawqa s-samᶜiyy, taḥta j-jildiyy, bayna l-qa:riyy*. However, our introduction in the questionnaire of constructions such as these, where independent prepositions are rendered into prefixes, has been prompted by our conviction that as a means of lexical creation this process represents a useful method of great potential in the language of science (see Chapter 3). Now let us examine the responses to these constructions.

As shown in Table 6, 33% were in favour of such constructions. This result constitutes a fair degree of significance. There is also the fact that 41% accepted these formations to a certain extent, which, is we take into consideration the respondents' unfamiliarity with them prior to the questionnaire, would suggest an even greater degree of potential for adaptability. It may therefore safely be argued that such constructions are likely to meet with a fair degree of success in scientific and technical language.

Coinages belonging to sample 3 were regarded as acceptable to a relatively lesser degree than those of samples 2 and 5. This can be accounted for by the fact that lexical formations involving the compounding of two or more words into one unit were not a prominent feature of Arabic. Such formations may thus be said to conflict with the general tendency of the language to use its words independently or, when necessary, in the form of set combinations of the types discussed above in 3.3.3 and 3.4. On the other hand, coinages of the present type (e.g. *nafsi:badaniyyu:n* "psychophysicists") were neverthless comparatively more acceptable than those in sample 4 (see below), and this may be because, from the point of view of their understandability, the former are far more transparent than the latter. The meaning of their invidual constituents, which are fully represented in their original morphological character, is not lost in the process of combination.

Regarding sample 4, the coinages introduced here were rejected by a significantly high percentage of the respondents: 86%. This result is in substantial agreement with the view expressed in Chapter 3 (3.3.3) that such coinages are "unlikely to achieve any success in actual linguistic practice". The reader may refer to the preceding pages of that section for an account of the arguments on which the opinion is based. Meanwhile, suffice it to say that the main factor behind the rejection of formations of this sort (e.g. *fatkalah* < *faṭhu l-kulyah* "nephrostomy") seems to be that they completely fail on the criterion of intelligibility since the meaning of their constituent elements is entirely lost in the process of abbreviation (Naḥt) and combination involved in their neologization; hence our designation of such coinages by the term "opaque N-constructions" (p.75).

Notice that as far as the number of their constituent sounds and their surface structure as lexical units are concerned, these coinages are in complete congruity with the typical character of Arabic native vocabulary. Nevertheless, this does not seem to make up for their failure on the criterion of intelligibility. Transparency may thus be described as an important factor in the acceptance of a term, though not the only one (cf. what has just been said with regard to sample 3 coinages).

Reference may also be made here to the general tendency of Arabic lexical words to relate to single roots as being another factor behind the opposition to coinages that are not so relatable. This same fact may be taken as an argument accounting for the favourable response to single-root formations of the type represented by sample 2, although the roots in this case are not of native origin.

Appendix A. Morphological patterning in Arabic: examples of analogically derived words

1 **Fu8a:L for names of diseases and ailments**

su^{c}a:l	cough
ṣuda:^{c}	headache
kuba:d	disease of the liver
ru^{c}a:f	bleeding (nose)
huza:l	emaciation
buwa:l	diabetes
xuna:q	diphtheria
zuma:l	disease of the joints
ruka:k	idiocy
ḥuma:q	smallpox
qula:^{c}	canker of the mouth
nuka:f	parotitis
juḥa:f	diarrhoea
duwa:r	giddiness
zuka:m	cold, catarrh
?ula:s	madness
buḥa:r	seasickness
hura:r	dysentery
sula:l	consumption
juwa:ḍ	melancholy
ḏuba:ḥ	croup
?uwa:r	thirst
jusa:d	bellyache
ṭuša:š	cold
fuqa:s	rheumatism
šuġa:f	pain in the pericardium
ḥula:q	sore throat
huya:m	passionate love
?uka:l	itch
ḏuba:l	tumour in the hypochondres
?uṭa:m	dysuria
mura:ḍ	mildew
qu^{c}a:ṣ	fatal disease of sheep from surfeit
kuta:f	pain in the shoulder

ḏu^{c}a:f	sudden death
ruma:^{c}	lumbago
zuḥa:r	hard breathing
^{c}uṭa:s	sneezing
fuwa:q	hiccough, gasp
suba:t	lethargy
luha:ṯ	panting, agony of death

2 **Fa8aL for names of physical defects, imperfections, diseases etc.**

ṣamam	deafness
xaras	dumbness
xazar	wryness of the eyes
ṣa^{c}ar	wryness of the face
^{c}awar	one-eyedness
baraṣ	white leprosy
namaš	freckles
ṣala^c	baldness
xaraf	dotage
ḥawal	squinting
wahan	weakness
saqam	disease
baḥaḥ	hoarseness
kasal	laziness
haram	decreptiude
^{c}araj	lameness
qazal	limping
xabal	insanity
ḥaban	dropsy
hawas	visionariness, foolishness
balah	stupidity
lamam	slight madness
^{c}amaš	blearedness
ramad	opthalmia
darad	the falling out of the teeth
qalaḥ	yellowness of the teeth
ḥafar	decayedness of the teeth (caries)
?araq	insomnia
qalaq	anxiety, agitation
waram	swelling, tumour
salas	incontinence of urine
xadar	numbness
sabal	cataract of the eye
maraḍ	illness
šalal	paralysis
wajal	fear
ta^{c}ab	fatigue
waja^c	pain
maġaṣ	gripes
ḍanaf	long illness
šajan	grief
ḏ̣ahar	backache
maḍaḍ	suffering, agony
jarab	scabies

3 Fa8aLa:n for nouns indicating motion, commotion and fluctuation

ṭayara:n	flying
sayala:n	flowing
fayaḍa:n	overflowing
jaraya:n	running, flowing
xafaqa:n	palpitation, fluttering
safaḥa:n	flowing of blood
ṯawara:n	agitation, excitement, eruption
wahaja:n	burning
wamaḍa:n	flashing
lamaᶜa:n	glowing
lahaba:n	blazing
rajafa:n	trembling
jawala:n	roving, wandering
rasafa:n	walking with the feet tied together
hadaja:n	tottering
katafa:n	stepping along quickly
xaṭara:n	swinging
tawaqa:n	longing with fervent desire
ġalaya:n	boiling
fawara:n	simmering
dawara:n	rotation
mayada:n	oscillation
zawala:n	passing away
ṣawala:n	attacking furiously
rawaġa:n	turning off
zayaġa:n	swerving
hayaja:n	commotion, agitation
ṭawafa:n	rambling, flooding
ṭayaha:n	straying, wandering
mayala:n	slanting, sloping

4 miF8aL, miF8a:L, miF8aLah for nouns of instrument

mišraṭ	lancet
mibḍaᶜ	penknife, lancet
miġzal	spindle
miṭmar	plumb-line
mibrad	file
miṯqab	borer, drill
mijrad	surgical instrument for cleaning the teeth
miᶜwal	pickaxe
mixšal	filtering funnel
miḥran	instrument for cleaning cotton
midᶜak	scourer
miqṭaᶜ	cutter
miqwad	halter, steering mechanism
milqaṭ	(pair of) tongs, tweezers
mijhar	microscope
midfaᶜ	cannon
mirqab	telescope
miṣᶜad	lift
misᶜar	calorimeter

mišbak	clasp
mišhaḏ̣	whetstone
mixraz	awl
misba:r	probe
miḍra:b	mallet, knocker
mifta:ḥ	key
miḥra:ṯ	plough
miḥra:k	poker, fire iron
mizla:j	latch
mirja:m	catapult
miqya:s	measuring instrument
mihra:s	mortar for pounding
minfa:x	bellows
miġwa:z	gasometer
mixma:r	vinometer
minġa:m	tonometer
minfa:s	spirometer
minqa:š	chisel
mij̲d̲a:f	oar
mizma:r	flute
mikya:l	measure
minḏ̣a:r	telescope
miṭya:f	spectroscope
miṭraqah	hammer
miṭḥanah	mill
mi^{c}zaqah	shovel
mizjafah	fishing-net
miġrafah	ladle, scoop
mindafah	teasing bow (for carding cotton)
mijrafah	trowel
miḥjamah	cupping-glass
mirmalah	sand-sifter
miknasah	broom
mil^{c}aqah	spoon
mimsaḥah	duster
minṭaqah	girdle
mi^{c}ṣarah	squeezer, oil press
mirjafah	tremograph
mirwaḥah	fan, propeller, ventilator
mišnaqah	gallows
miqṣalah	guillotine

Appendix B. Resolutions passed by the Cairene Academy concerning lexical creation in Standard Arabic

The following are a number of principles and resolutions passed and adopted by the Arabic Language Academy of Cairo by way of regulating the process of creating new vocabulary by means of native-based methods, especially analogical derivation. They are listed here for the benefit of translators and people who are concerned with the problem of lexical deficiency in Arabic. Examples are also provided.

(i) Nouns of instrument

Derivatives are to be analogically formed from triradical verbs according to the patterns miF8aL, miF8a:L, miF8aLah for the denotation of an instrument by which something is manipulated. For example: *mibrad* "file", *miṭya:f*, "spectroscope", *miᶜṣarah* "squeezer" [for more examples, see Appendix A, 4 above]. The Academy also decrees that the patterns of instrumental nouns [already] in common use be adhered to. [Such non-analogical patterns include muF8uL, e.g. *munxul* "sieve"; Fi8a:L, e.g. *?ira:ṯ* "instrument for kindling fire"; Fu8L, e.g. *mušṭ* "a comb", etc.] [Otherwise] if no such pattern has been formed from a verb, this may be performed in accordance with any of the three patterns previously mentioned.

(ii) Derivation from concrete nouns

The [early] Arabs frequently used concrete nouns for further derivation; [therefore] the Academy approves this [type of] derivation, in case of necessity, in the language of science, for example: *mukarban* "carbonated" < "carbon"; *tamaġnuṭ* "magnetization" < "magnet", etc.

(iii) Nouns of abstraction

A noun of abstraction may be derived from a word by terminating it by [the morpheme] *-iyyah*. Examples: *lawniyyah* "chromaticity", *magna:ṭi:siyyah* "magnetism", *?iḥtira:qiyyah* "combustibility", etc.

(iv) Nouns of profession, activity, post, etc.

An abstract noun of the pattern Fi8a:Lah is to be formed from any category of the triliteral [i.e. whether the corresponding verb is of the form Fa8aL-, Fa8iL- or Fa8uL-] to denote a profession or something that approximates to it. For example: *nija:rah* "carpentry", *jira:ḥah* "surgery", *qiya:dah* "leadership", etc.

(v) Nouns of motion, commotion, fluctuation, etc.

An abstract noun of the pattern Fa8aLa:n is analogically applicable in the case of intransitive verbs of the form Fa8aL-, when denoting fluctuation or commotion (*taqallub ?aw ?iḍṭira:b*). [What is implied here is any type of movement involving agitation, vibration, convulsion, oscillation, swinging, trembling, etc.] For examples, see Appendix A, 3.

(vi) The pattern Fu8a:L for names of diseases

An abstract noun of the pattern Fu8a:L denoting a disease is to be analogically derived from an intransitive verb of the form Fa8aL-. For example: *suᶜa:l* "cough" < *saᶜal-* "to cough".

(vii) Nouns denoting sounds

If the language contains no abstract noun corresponding to an intransitive verb of the type Fa8aL- which denotes a sound, such a noun is to be analogically formed according to the pattern Fu8a:L or Fa8i:L. For example: from *nabaḥ-* "to bark" we get *nuba:ḥ* or *nabi:ḥ* "barking".

(viii) The pattern Fa88a:L for indicating reference or relation

The pattern Fa88a:L is to be analogically formed to denote professional pursuit of a trade, etc. (*?iḥtira:f*) or (mere) association (*mula:zamah*) to something. In cases where ambiguity might arise between the maker (producer or manufacturer) of something and the one that is associated with it, the pattern Fa88a:L should be used with the former, and the Nisba-adjective (i.e. *-iyy*) with the latter. For example: *zajja:j* is "a maker or producer of *zuja:j* "glass", and *zuja:jiyy*, "one who sells it".

(ix) Transitivization by means of al-hamzah, i.e. /?a-/

The Academy regards as analogically applicable the transitivization of the intransitive triliteral verb by [initiating it with] the *hamzah*. For example: *qa:m-* "to stand up" → *?aqa:m-* "to make somebody rise; to erect".

(x) Analogical formation of verbs of consequence or compliance

?afᶜa:lu l-muṭa:waᶜah; other terms used to designate the same type of verbs include "quasi-active", "quasi-passive" and "reflexive".

(a) Fa8aL- → ?inFa8aL-: Any transitive triliteral verb of the form Fa8aL- which denotes physical manipulation (or processing, etc.) has the form ?inFa8aL- as its analogical verb of consequence unless the first radical (i.e. of the former) is a /w/ /l/ /n/ /m/ or /r/, in which case the corresponding analogical form would be ?iFta8aL-. For example: *kasar-* "to break" → *?inkasar-* "to get (or be) broken"; but *mazaj-* "to mix" → *?imtazaj-* "to be mixed.

(b) Fa88aL- → taFa88aL-: e.g. *farraq-* "to scatter" → *tafarraq-* "to be scattered"; *ᶜallam-* "to teach" → *taᶜallam-* "to learn".

(c) Fa:8aL- → taFa:8aL-: e.g. *wa:zan-* "to equal in weight" → *tawa:zan-* "to be equal in weight, to be in equilibrium".

(d) Fa8LaL- → taFa8LaL-: e.g. *daḥraj-* "to roll" → *tadaḥraj-* "to roll along, roll down"; *jalbab-* "to dress with a *jilba:b* "wide shirt → *tajalbab-* "to be dressed with it".

(xi) The pattern ?istaF8aL- to express the meaning of "to request" or "to become" (liṭ-ṭalabi wa ṣ-ṣayru:rah)

The Academy deems that the pattern ?istaF8aL- is analogical for indicating (the concept of) request or (the process of) becoming. For example *?istaġfar-* "to ask (somebody's) pardon, to ask (somebody) to forgive"; cf. *ġafar-* "to forgive"; *?istaḥjar-* "to turn into stone *ḥajar*"; *?ista?sad-* "to display the courage of a lion *?asad*", etc.

(xii) The pattern Fa88aL- for intensiveness

e.g. *kassar-* "to fragmentize" < *kasar-* "to break"; *qattal-* "to massacre" < *qatal-* "to kill"; *jarraḥ-* "to inflict many and severe wounds" < *jaraḥ-* "to wound".

(xiii) Relative adjectives from plural forms

A relative adjective may be formed from a plural form when this is seen necessary, as in the case of distinction, e.g.: *?axla:qiyy* "moral" < *?axla:q* "manners"; *suwariyy* "formal, artificial, imaginary" < *ṣuwar* "pictures, forms"; *waṯa:?iqiyy* "documentary, factual, objective" < *waṯa:?iq* "documents".

Appendix C. Examples of Full (Opaque) Naht-constructions

N-construction	Set combination involving the underlying constituents
nazkarah (decarbonization)	*nazᶜu l-karbo:n*
naz-warah (defoliation)	*nazᶜu l-waraq*
wajᶜadah (gastralgia)	*wajaᶜu (ʔalamu) l-maᶜidah*
fatmanah (cystostomy)	*fatḥu l-maṯa:nah*
fatkalah (nephrostomy)	*fatḥu l-kulyah*
ṣalkalah (nephrectomy)	*ʔistiʔṣa:lu l-kulyah*
qaṭmaᶜah (enterectomy)	*qaṭᶜu l-ʔamᶜa:ʔ*
šibza:l (albuminoid)	*šibhu z-zula:l*
ḥarṣam- (to de-gum)	*ḥarrara mina ṣ-ṣamġ*
zaḥrajah (deforestation)	*ʔiza:latu l-ʔaḥra:j*
ḥargaz- (to de-gas)	*ḥarrara mina l-ġa:z*
qaṭjaraḥ (laryngotomy)	*qaṭᶜu l-ḥanjarah*
xazladah (oxidation-reduction)	*al-ʔaksadatu wa l-ʔixtiza:l*
ḥawbaġ (zoospore)	*al-bawġu l-ḥayawa:niyy*
ḥarkabiyy (thermoelectric)	*kahraba:ʔiyyun ḥara:riyy*
naqrab- (to electro-engrave)	*naqasa bil-kahraba:ʔ*

kahramya:? (electrochemistry)	*al-ki:mya:?u l-kahraba:?iyyah*
ḥalqaḍah (daydream)	*ḥulmu l-yaqḍah*
sarnamah (somnambulism)	*as-sayru xila:la n-nawm*
li^{c}nafiyya:t (malacopterygii)	*layyina:tu z-za^{c}a:nif*
an-nazjanah (dehydrogenation)	*naz^{c}u l-hi:dro:ji:n*
al-ḥarkabah (electromotion)	*al-ḥarakatu l-kahraba:?iyyah*
al-karṣalah (polygenesis)	*kaṯratu l-?uṣu:l*
al-xalḍa?ah (photosynthesis)	*at-taxli:qu ḍ-ḍaw?iyy*
aḍ-ḍawkar (photosphere)	*al-kuratu ḍ-ḍaw?iyyah*
at-tama:kub (isomerism)	*at-tama:ṯulu bit-tarki:b*
al-baṭjal (gastropod)	*al-baṭniyyu l-?arjul*
aṯ-ṭarḍa:w?iyyah (photosensitivity)	*at-ta?aṯṯuru bin-naša:ṭati ḍ-ḍaw?iyyah*

Appendix D. Partial (Semi-transparent) Naht-constructions

N-construction	**Set combination involving underlying constituents**
musjana:ḥiyya:t (orthoptera)	*mustaqi:ma:tu l-?ajniḥah*
ᶜaṣjana:ḥiyy (neuropterous)	*ᶜaṣabiyyu l-?ajniḥah*
az-zamaka:n (space-time)	*az-zama:n - al-maka:n*
nafjismiyy (psychophysical)	*nafsiyyun jismiyy*
niṣšawkiyy (semi-spinal)	*niṣfu š-šawkiyy*
jaḏra?siyya:t (rhizocephala)	*jaḏriyya:tu l-?ar?us*
juzmusannan (subdental)	*musannanun juz?iyyan*
az-zayšawkiyy (olivo-spinal)	*az-zaytu:niyyu š-šawkiyy*
al-ᶜay?anfiyy (oculo-nasal)	*al-ᶜayniyyu l-?anfiyy*
?arbayad (quadrumane)	*ruba:ᶜiyyu l-?aydi:*
ṯilra?siyy (tricuspid)	*ṯula:ṯiyyu r-ra?s, ḏu: ṯala:ṯati ru?u:s*
ṯinlawniyy (dichromatic)	*ṯuna:?iyyu l-lawn, ḏu: lawnayn*
šibġarawiyy (colloidal)	*šibhu ġarawiyy*

Appendix E. Arabic prepositions as prefixes

Pref.	Prep.	N-construction	
qab-	< *qabla*	*qabmada:riyy*	preorbital
		qabṭa:ḥin	premolar
		qabmura:haqah	preadolescence
ġib-	< *gibba*	*ġibmadrasiyy*	postscholastic
		ġibjali:diyy	postglacial
		ġibḥarbiyy	postwar
faw-	< *fawqa*	*fawbašariyy*	superhuman
		fawqawmiyy	supernational
		fawᶜuḍwiyy	superorganic
taḥ-	< *taḥta*	*taḥjildiyy*	subcutaneous
		taḥmarkaziyy	subcentral
		taḥfakkiyy	submaxillary
du:-	< *du:na*	*du:mijhariyy*	submicroscopic
		du:šamsiyy	subsolar
		du:ḏarriyy	subatomic
bay-	< *bayna*	*bayqa:rriyy*	intercontinental
		bayfiṣṣiyy	interlobular
		baykawkabiyy	interplanetary
ḥaw-	< *ḥawla*	*ḥawšamsiyy*	circumsolar
		ḥawquṭbiyy	circumpolar
		ḥawqamariyy	circumlunar
xal-	< *xalfa*	*xalmiḥwariyy*	postaxial
		xal?anfiyy	postnasal
		xalqalbiyy	postcardinal
ḍim-	< *ḍimna*	*ḍimnafsiyy*	intrapsychic
		ḍimxalawiyy	intracellular
		ḍimnawᶜiyy	intraspecific

Appendix F. Examples of modern terms and arabicized words

The examples given below are quoted from a number of Arabic articles and books dealing with scientific topics. Here is first a list of the sources where the contextual occurrences of these examples have been found:

a Majīd al-Qaysiyy, "Dirāsatun fī l-Kīmyā? al- Ishᶜāᶜiyyah", in *al-Ustādh* ("The Professor", an academic review issued by the College of Education, University of Baghdad), vol.xii, 1963-4, pp.381-9.

b Iḥsān al-ᶜĪsa, "ᶜIlāqat al- Ibdāᶜ bi-ᶜUmur Marḍa al-Fuṣām", *al-Ustādh*, vol.xii, 1963-4, pp.390-402.

c Murād Bābā Murād (trans.), " Ighrāq al-Mawādd al-Mushiᶜᶜah fī al-Baḥr", *al-Ustādh*, vol.xii, 1963-4, pp.422-31.

d Salīm ᶜAbd al-Ḥamīd Qāsim, "Muᶜāmil al- Imtiṣāṣ al-Kutaliyy wa ᶜAddād al-ṭala?lu?", *al-Ustādh*, vol.xiii, 1964-5, pp.330-5.

e Jalāl Muḥammad Ṣāliḥ, Aḍwā?un Ḥadīthah ᶜala al-Imtiṣāṣ al-Kīmyāwiyy", *al-Ustādh*, vol.xiv, 1066-7, pp.383-418.

f ᶜAlwān al-Wā?iliyy, "al-Khaliyyah Taḥt al-Mīkrōskōb al-Ilikitrōniyy", *al-Ustādh*, vol.xiv, 1966-7, pp.482-92.

g ᶜĀdil Saᶜīd Waṣfi, " al-Judhūr al-Ḥurrah", *al-Ustādh*, vol.xiv, 1966-7, pp.419-47.

h Ibrāhīm Yūsuf al-Manṣūr, "Mafhūm al-Qānūn al-Ṭabīᶜiyy fī ᶜIlm al-Nafs", in *Majalat Kulliyyat al-Adab* (Bulletin of the Faculty of Arts, University of Baghdad), vol.xiii, 1965-6, pp.121-37.

i Ṣabīḥ al-Ḥāfiḍh, "Mā huwa al-Maikrōfilm", Baghdad, Dār al-Rashīd lil-Nashr, 1979.

j ᶜAbd al-Muḥsin Ṣāliḥ, "Ṭabkhatun Arḍiyyah Tumahhid li-Ādam al-Juzayʔāt al-Ḥayyah", in *al-ᶜArabiyy*, no.253, December 1979, pp.52-7.

k ᶜAbd al-Bāqī al-Ṣāfī, "Dirāsatun Maqāranah lil-Kalimah wa ᶜIlm al-Ṣarf fī al-Lughatayn al-ᶜArabiyyah wa al-Injilīziyyah", in *al-Mirbad* (Bulletin of the Faculty of Arts, University of Basrah, Iraq), no.4, 1970, pp.173-218.

mula?li?	scintillator	[d p332]
munaḏḏimu l-?iḍa:?ah	exposure knob	[i p 93]
ᶜilmu ṣ-ṣana:?iᶜ	technology	[c p430]
al-mawa:ddu šibhu l-muwaṣṣilati d-da:xiliyyah	intrinsic semiconductor	[e p392]
as-sa:libiyyati l-kahraba:?iyyah	electronegativity	[e p397]
al-ᶜaṣi:ru n-nawawiyy	nucleoplasm	[f p486]
al-fara:ġu l-muḥi:ṭu bin-nuwwa:h	perinuclear space	[f p488]
al-?afla:m, ?u:tu:ma:ti:kiyyah, al-muḏḏahhir	films, automatic, developer	[i p 81]
at-tawṣi:lu l-?amṯal	superconductivity	[g p435]
at-ta?ṯi:ra:tu l-kahraba:?iyyatu l-ḥara:riyyah	thermoelectric effects	[g p444]
al-ḥa:siba:t	computers	[g p445-6]
al-qiya:su n-nafsiyy	psychometrics	[h p134]
ᶜilmu l-xilyah	cytology	[f p482]
ᶜilmu n-nafs	psychology	[h p121]
sa:yku:lu:jiyy	psychological	[h p121]
al-?imtiṣa:ṣ	adsorption	[e p383]
al-kahrawḍaw?iyyah	photoelectric	[g p444]
al-ma:kro:film	microfilm	[i p 16]
sama:ḥiyyatu l-maġna:ṭi:siyyah	magnetic permitivity	[g p445]
fawqu l-banafsajiyyah	ultraviolet	[g p424]
at-taḥli:lu l-kahraba:?iyy	electrolysis	[g p425]
siya:ni:du l-ḥadi:di:k	ferricyanide	[g p419]
al-ba:ḥiṯi:nu l-biyu:lu:jiyyi:n	biologists	[f p482]
?aksadah	oxidization	[e p401]
?aka:si:d	oxides	[e p400]
al-haljanah	halogenation	[g p428]
al-faslajah	physiology	[h p121]
al-ġa:za:tu l-muta?ayyanah	ionized gases	[g p446-7]
ta?yi:n hi:dru:ji:n	ionization hydrogen	[c p395]
al-balmarah	polymerization	[g p420]
al-mi:ṯa:n, al-?i:ṯa:n, ṯa:ni: ?u:ksi:du l-ka:rbu:n, xilla:tu l-mi:ṯi:l	methane, ethane, carbon dioxide, methyl acetate	[g p422]
buru:m ha:ydru:ka:rbu:r	bromine hydrocarbon	[g p420]
al-mi:kru:sku:bu l-?iliktru:niyy	electron microscope	[f p483]
fi:ru:maġna:ṭi:siyyah	ferromagnetic	[g p437]
?antifi:ru:maġna:ṭi:siyyah	antiferromagnetic	[g p437]
at-taḥli:lu l-kru:mu:tu:ġra:fiyy	chromotographic analysis	[e p383]
aš-ši:zu:fari:niyyah	schizophrenia	[b p390]
mi:kru:ġara:ma:, ġara:m, yu:ra:nyu:m	microgram, gram, uranium	[a p382]

Bibliography

This bibliography consists not only of the works cited and directly quoted in the text, but also of books which are useful for those interested in the study and development of Arabic vocabulary.

For the sake of convenience, some periodicals are cited in an abbreviated form:

MM^{c}I^{c}I	=	Majallat al-Majma^{c} al-^{c}Ilmiyy al-^{c}Irāqiyy (Journal of the Iraqi Academy)
MM^{c}I^{c}A	=	Majallat al-Majma^{c} al-^{c}Ilmiyy al-^{c}Arabiyy (Journal of the Syrian Academy)
MML^{c}AM	=	Majallat Majma^{c} al-Lughah al-^{c}Arabiyyah al-Malakiyy (Journal of the Royal Academy of the Arabic Language, Cairo)
MML^{c}A	=	Majallat Majma^{c} al-Lughah al-^{c}Arabiyyah (Journal of the Arabic Language Academy, Cairo)

^{c}Abd al-raḥmān, Ḥikmat Najīb (1977)
"Dirāsāt fī Ta?rīkh al-^{c}Ulūm ^{c}ind al-^{c}Arab", Iraq.
^{c}Abd al-Raḥmān, ^{c}Ā?ishah (1971)
"Lughatuna wa l-Ḥayāt", Cairo.
Abu al-Faraj, Muḥammad Aḥmad (1961)
"al-Ma^{c}ājim al-Lughawiyyah", Dār al-Nahḍah al-^{c}Arabiyyah.
Adams, Valerie (1973)
"An Introduction to Modern English Word Formation", London.
al-Afghāniyy, Sa^{c}īd (1936)
"Kalimat Ḥiyād", MM^{c}IA, vol.xiv, no.4, pp.147-52.
al-Afghāniyy, Sa^{c}īd (1969)
"al-Mūjaz fī Qawā^{c}id al-Lughah al-^{c}Arabiyyah", Beirut.
al-^{c}Alāyli, ^{c}Abd Allah (1940)
"Muqaddamah li-Dars Lughat al-^{c}Arab", Cairo.

Anīs, Ibrāhīm (1950)
"Al-Aswāṭ al-Lughawiyyah", 2nd edn, Cairo.
Anīs, Ibrāhīm (1950)
"Min Asrār al-Lughah", 2nd edn, Cairo.
Anīs, Ibrāhīm (1976)
"Dalālat al-Alfāḍh" 3rd edn, Cairo.
Arabic Language Academy (Cairene Academy) (1957)
"Majmūᶜat al-Muṣṭalaḥāt al-ᶜIlmiyyah wa al-Fanniyyah allatī aqarraha al-Majmaᶜ", Cairo.
al-Athariyy, M.B. (1963)
"al-Ālah wa al-Adāt fī al-Lughah al-ᶜArabiyyah", MMᶜIᶜI, vol.x, pp.4-29.
Bakalla, Muhammad Hasan (1970)
"The Phonetics and Phonology of Classical Arabic as Described in Ibn Jinni's Sirr al-Ṣināᶜah", PhD dissertation, London, SOAS.
al-Baᶜlabakkiyy, Munir (1973)
"al-Mawrid", 6th edn, Beirut.
Bāqir, Ṭāha (1973)
"Muqaddamah fī Ta?rīkh al-Ḥaḍārāt al-Qadīmah", Baghdad.
Beeston, A.F.L. (1970)
"The Arabic Language Today", London.
Blochet, E. (1896)
"Note sur l'arabisation des mots persans", Revue Sémitique d'Epigraphie et d'Histoire Ancienne, Recueil Trimestriel, Paris, 4e annee, pp.266-72.
Bloomfield, Leonard (1973)
"Language", London.
Bradley, H. (1920)
"The Making of English", London.
Browne, Edward G. (1921)
"Arabian Medicine", Cambridge.
al-Bustāniyy, Buṭrus (1870)
"Muḥīṭ al-Muḥīṭ", 2 vols, Beirut.
Cantineau, J. (1950)
"La notion de 'schème' et son altération dans diverses langues sémitiques", Semitica, III, pp.73-83.
Chejne, Anwar G. (1969)
"The Arabic Language, Its Role in History", Minnesota.
Darwīsh, A.A.F. (1955)
"al-Khalīl Ibn Aḥmad and the Evolution of Arabic Lexicography", 2 vols, PhD dissertation, London, SOAS.
Dozy, Reinhart (1913)
"Spanish Islam: A History of the Moslems in Spain", London.
Erwin, Wallace M. (1963)
"A Short Reference Grammar of Iraqi Arabic", Georgetown.
Fahmi, Ḥasan Ḥusayn (1961)
"Al-Marjiᶜ fī Taᶜrīb al-Muṣṭalaḥāt al-ᶜIlmiyyah wa al-Fanniyyah wa al-Handasiyyah", Cairo.
al-Fārābiyy, Muḥammad Abū Naṣr (1931)
"Iḥṣā? al-ᶜUlūm", ed. ᶜUthmān Muḥammad Amīn, Cairo.
al-Fayrūzābādiyy, Muḥammad bin Yaᶜqūb (1301-3 AH)
"al-Qāmūs al-Muhīt", 4 vols, 3rd edn, ed. Naṣr al-Hūrīniyy, Cairo.

Ferguson, C.A. (1970)
"Arabic Language", Encyclopaedia Britannica, vol.2, Chicago edn, pp.182-4.
Fleisch, Henri (1956)
"L'Arabe classique: Esquisse d'une structure linguistique", Beirut.
Ghazāl, Aḥmad al-Akhḍar (1977)
"al-Manhajiyyah al-ᶜĀmmah li-Taᶜrīb al-Muwākib", Rabat.
Ghazāl, Aḥmad al-Akhḍar (1978)
"Uslūb Ikhtiyār al-Muṣṭalaḥ al-ᶜIlmiyy", paper presented to the Conference of Arabization, Baghdad.
Ghazāl, Aḥmad al-Akhḍar (n.d.)
"al-Manhajiyyah al-Jadīdah li-Waḍᶜ al-Muṣṭalaḥāt al-ᶜArabiyyah", Rabat.
Gibb, H.A.R. (1963)
"Arabic Literature", 2nd edn, Oxford.
Goichon, A.M. (1969)
"The Philosophy of Avicenna and its Influence on Medieval Europe", Motilal Banarsidass, India.
Greenberg, Joseph H. (1950)
"The Patterning of Root Morphemes in Semitic", Word, VI, 2.
Greenberg, Joseph H. (1957)
"Essays in Linguistics", Chicago.
Ghuṣn, Mārūn (1935)
"al-Naḥt fī al-Lughah al-ᶜArabiyyah Wasīlah li-Tawsīᶜ al-Lughah", MMᶜIᶜA, vol.xiii, no.6, pp.300-302.
Hall, Robert A. (1964)
"Introductory Linguistics", Philadelphia.
al-Ḥarīriyy, Abū Muḥammad al-Qāsim bin ᶜAli (1299 AH)
"Durrat al-Ghawwāṣ fī Awhām al-Khawāṣṣ", Constantinople.
Ḥasan, ᶜAbbās (1963)
"al-Naḥw al-Wāfi", 4 vols, Cairo.
Ḥassān, Tammām (1973)
"al-Lughah al-ᶜArabiyyah, Maᶜnāha wa Mabnāha", Cairo.
Hassān, Tammām (1974)
"Manāhij al-Baḥth fī al-Lughah", 2nd edn, al-Dār al-Bayḍā?.
Haugen, Einar (1950)
"The Analysis of Linguistic Borrowing", Language XXVI, pp.210-31.
Ḥijāziyy, M.F. (1970)
"ᶜIlm al-Lughah bayn al-Turāth wa al-Manāhij al-Ḥadīthah", Cairo.
Ḥijāziyy, M.F. (1978)
"al-Lughah al-ᶜArabiyyah ᶜAbr al-Qurūn", Cairo.
Hoijer, Harry (1964)
Linguistic and Cultural Change, in D. Hymes, "Language in Culture and Society: a reader in linguistics and anthropology", New York, pp.455-66.
Ḥusayn, Muḥammad al-Khiḍr (1960)
"Dirāsāt fī al-ᶜArabiyyah wa Ta?rīkhiha", 2nd edn, Damascus.
al-Ḥusayniyy, Abū al-Baqā? (1974-76)
"al-Kulliyyāt", 5 vols, eds ᶜAdnān Darwīsh and Muḥammad al-Miṣriyy, Damascus.

al-Ḥuṣriyy, Sāṭiᶜ (1958)
"Ārā? wa Aḥādīth fī al-Lughah wa al-Adab", Beirut.
Ibn Abdallah, Abdel-Aziz (1976)
"Problems of Arabization in Science", al-Lisān al-ᶜArabiyy, vol.xxvi, no.3, Rabat, pp.v-xiii.
Ibn al-Anbāriyy (1364 AH)
"Kitāb al-Inṣāf fī Masā?il al-Khilāf bayn al-Naḥwiyyīn al-Baṣriyyīn wa al-Kūfiyyīn", Cairo.
Ibn ᶜArabiyy, Muḥyi al-Dīn (1972-78)
"al-Futūḥāt al-Makkiyyah", 6 vols, ed. ᶜUthmān Yaḥya with preface and revision by Ibrāhīm Madhkūr, Cairo.
Ibn Durayd, Muḥammad bin al-Ḥasan (1958)
"al-Ishtiqāq", ed. ᶜAbd al-Salām Muḥammad Hārūn, Maṭbaᶜat al-Sunnah al-Muḥammadiyyah.
Ibn Durayd, Muḥammad bin al-Ḥasan (1344-45 AH)
"Jamharat al-Lughah", 3 vols, Hyderabad.
Ibn Fāris (1963)
"al-Ṣāḥibiyy fī Fiqh al-Lughah wa Sunan al-ᶜArab fī Kalāmiha", ed. Muṣṭafa al-Shuwaymiyy, Beirut.
Ibn Fāris (1946-51)
"Muᶜjam Maqāyīs al-Lughah", 6 vols, ed. ᶜAbd al-Salām Muḥammad Hārūn, Cairo.
Ibn Jinniyy, ᶜUthmān (1913)
"al-Khaṣā?iṣ", Cairo.
Ibn Jinniyy, ᶜUthmān (1954)
"Sirr Ṣināᶜat al-Iᶜrāb" eds Muṣṭafa al-Saqqā et al., Cairo.
Ibn Manḍhūr (1955-56)
"Lisān al-ᶜArab", 15 vols, Beirut.
Ibn Qutaybah, Abu Muḥammad (1300 AH)
"Adab al-Kātib", Cairo.
Ibn Sīdah (n.d.)
"al-Mukhaṣṣaṣ", 5vols, Beirut.
Iraqi Academy (1967)
"Muṣṭalaḥāt Muqāwamat al-Mawādd wa Handasat al-Mā?", Baghdad.
Iraqi Academy (1977)
"Muᶜjam Muṣṭalaḥāt al-Fīzyā?", Baghdad.
al-Iskandariyy, Ahmad (1935)
"al-Gharaḍ min Qarārāt al-Majmaᶜ wa al-Iḥtijāj laha", MMLᶜAM, vol.i, pp.177-268.
al-Jāḥiḍh, Abūᶜ Uthmān ᶜAmr b. Baḥr (1961)
"al-Bayān wa al-Tabyīn", 4 vols, 2nd edn, ed. ᶜAbd al-Salām M. Hārūn, Cairo.
Jawād, Muṣṭafa (1955)
"al-Mabāḥith al-Lughawiyyah fī al-ᶜIrāq", Cairo.
Jawad, Muṣṭafa (1957)
"Wasā?il al-Nuhūḍ bil-Lughah al-ᶜArabiyyah", MMᶜIᶜA, vol.xxxii, pp.129-60.
al-Jawāliqiyy, Mawhūb bin Ahmad (1969)
"al-Muᶜarrab min al-Kalām al-Aᶜjamiyy", 2nd edn, ed. Aḥmad Muḥammad Shākir, Cairo.
al-Jawhariyy, Ismāᶜīl bin Hammād (1375-77 AH)
"al-Ṣiḥāḥ Tāj al-Lughah wa Ṣiḥāḥ al-ᶜArabiyyah", 6 vols, ed. Aḥmad ᶜAbd al-Ghafūr ᶜAṭṭār, Cairo.

Jeffery, Arthur (1938)
"The Foreign Vocabulary of the Qur?ān", Baroda.
Jesperson, Otto (1922)
"Language: Its Nature, Development and Origin", London.
Jirjis, Ramsīs (1961)
"al-Naḥt fī al-^{c}Arabiyyah", MML^{c}A, vol.xiii, 1961, pp.61-76.
al-Jundī, Salīm (1935)
"Mā Hākadha Yā Sa^{c}d Tūrad al-Ibil", MM^{c}I^{c}A, vol.xiii, no.8, pp.359-62.
al-Jurjāniyy, ^{c}Ali bin Muḥammad (al-Sharīf) (1357 AH)
"al-Ta^{c}rīfāt", Cairo.
al-Karmiyy, Sa^{c}īd (1921)
"al-Lughah wa al-Dakhīl fīha", MM^{c}I^{c}A, vol.i, no.5, pp.129-37.
al-Kasimi, Ali (1978)
"Problems of technical terminology in Arabic lexicography", al-Lisān al-^{c}Arabiyy, vol.xvi, no.1, pp.15-24.
al-Kawākibiyy, Ṣalāḥ al-Dīn (1947)
"Muṣṭalaḥāt ^{c}Ilmiyyah", 3rd edn, Damascus.
al-Karmaliyy, Anastās Mārī (1938)
"Nushū? al-Lughah al-^{c}Arabiyyah wa Numuwwuha wa Iktihāluha", Cairo.
al-Khafājiyy, Shihāb al-Dīn Aḥmad (1325 AH)
"Shifā? al-Ghalīl fīma fī Kalām al-^{c}Arab min al-Dakhīl", Cairo.
al-Khalīl bin Aḥmad (1967)
"Kitāb al-^{c}Ayn", vol.i, ed. A.A. Darwīsh, Baghdad.
al-Khaṭīb, Aḥmad Shafīq (1978)
"A New Dictionary of Scientific and Technical Terms", 4th edn, Beirut.
al-Khuwārizmiyy, Abū ^{c}Abdallah (1895)
"Mafātīḥ al-^{c}ulūm", ed. G. van Vloten, Leyden.
Ladefoged, Peter (1975)
"A Course in Phonetics", New York.
Lughat al-^{c}Arab (1911-)
Founded and edited by Anastās Mārī al-Karmaliyy, Baghdad.
Maḍhhar, Ismā^{c}īl (n.d.)
"Tajdīd al-^{c}Arabiyyah", Cairo.
Maḍhhar, Ismā^{c}īl (n.d.)
"al-Nahḍah Dictionary", 2 vols., Cairo.
al-Maghribiyy, ^{c}Abd al-Qādir (1947)
"Kitāb al-Ishtiqāq wa al-Ta^{c}rīb", 2nd edn, Cairo.
Majallat al-Majma^{c} al-^{c}Ilmiyy al-^{c}Arabiyy (MM^{c}I^{c}C) (1921-)
vols.1-, Damascus.
Majallat al-Majma^{c} al-^{c}Ilmiyy al-^{c}Irāqiyy (MM^{c}I^{c}I) (1950-)
Baghdad.
Majallat Majma^{c} al-Lughah al-^{c}Arabiyyah al-Malakiyy (MML^{c}AM) (1935-57) vols.1-3, Cairo.
Majallat Majma^{c} Fu?ād al-Awwal lil-Lughah al-^{c}Arabiyyah (1939-51) vols.4-6, Cairo.
Majallat Majma^{c} al-Lughah al-^{c}Arabiyyah (1953-)
vols.7-, Cairo.
Ma^{c}lūf, Luwīs (1966)
"al-Munjid", 19th edn, Beirut.

Marchand, H. (1960)
"The Categories and Types of Present-Day English Word Formation", Wiesbaden.

Matthews, C.M. (1979)
"Words", London.

al-Makhzūmiyy, Mahdī (1955)
"Madrasat al-Kūfah wa Manāhijuha fī Dirāsat al-Lughah wa al-Naḥw", Baghdad.

al-Makhzūmiyy, Mahdī (1974)
al-Dars al-Naḥwiyy fī Baghdād", Baghdad.

Mazhar, M.A. (1963)
"Arabic: The Source of all the Languages", Lahore.

Mitchell, T.F. (1960)
"Prominence and syllabication in Arabic", BSOAS, vol.xxiii, pp.369-89.

al-Mubārak, Muḥammad (1960)
"Fiqh al-Lughah", Damascus.

Muḥammad, ᶜAbd al-ᶜAzīz, et al. (1965)
"Muᶜjam al-Muṣṭalaḥāt al-ᶜIlmiyyah", Cairo.

Nasr, Seyyed Hossein (1968)
"Science and Civilisation in Islam", Cambridge, Mass.

Nicholson, Reynold A. (1953)
"A Literary History of the Arabs", Cambridge.

O'Connor, J.D. (1974)
"Phonetics", Harmondsworth.

O'Leary, De Lacy (1949)
"How Greek Sciences Passed to the Arabs", London.

O'Leary, De Lacy (1954)
"Arabic Thought and Its Place in History", rev. edn, London.

Onions, C.T. (1966)
"The Oxford Dictionary of English Etymology", ed. with G.W.S. Friedrichsen and R.W. Burchfield), Oxford.

Palmer, F.R. (1971)
"Grammar", London.

Pei, Mario (1966)
"Glossary of Linguistic Terminology", New York.

The Permanent Bureau of Arabization (1971)
"Lexicon of Chemistry", Rabat.

al-Rāziyy, Abū Bakr (1927)
"al-Madkhal al-Taᶜlīmiyy", in H.E. Stapleton et al. "Chemistry in ᶜIraq and Persia in the Tenth Century AD", Memoires of the Asiatic Society of Bengal, Calcutta, vol.xiii, no.6, pp.412-17.

al-Rāziyy, Abū Bakr (1964)
"al-Asrar wa Sirr al-Asrār", ed. Muhammad Taghi Danechpajouh, Tehran.

al-Rāziyy, Muḥammad bin ᶜUmar (Fakhr al-Dīn) (1343 AH)
"al-Mabāḥith al-Mashriqiyyah", 2 vols, Hyderabad.

Robins, R.H. (1975)
"General Linguistics: An Introductory Survey", 2nd edn, London.

al-Ṣāfī, ᶜAbd al-Bāqī (1970)
"Dirāsah Muqāranah lil-Kalimah wa ᶜIlm al-Ṣarf fī l-Lughatayn al-ᶜArabiyyah wa al-Injilīziyyah", Al-Mirbad (Bulletin of the Faculty of Arts, Basrah University, Iraq), no.4, pp.173-218.

al-Sāmarrā?iyy, Ibrāhīm (1977)
"al-Lughah wa al-Ḥaḍārah", Beirut.
Sapir, E. (1921)
"Language: An Introduction to the Study of Speech", London.
Savory, H. (1953)
"The Language of Science", London.
al-Sayyid, Amīn ʿAli (1976)
"Fī ʿIlm al-Ṣarf", 3rd edn, Cairo.
Sheard, J.A. (1954)
"The Words We Use", London.
Sībawayh (1361 AH)
"Kitāb Sībawayh", 2 vols, Bulaq.
al-Shihābiyy, Muṣṭafa (1955)
"al-Muṣṭalaḥāt al-ʿIlmiyyah fī al-Lughah al-ʿArabiyyah", Jāmiʿat al-Duwal al-ʿArabiyyah Maʿhad al-Dirāsāt al-ʿArabiyyah al-ʿĀli.
al-Shihābiyy, Muṣṭafa (1959)
"Madā al-Naḥt fī al-ʿArabiyyah", MMʿIʿA, vol.xxxiv, no.4, pp.545-54.
al-Shihābiyy, Muṣṭafa (1965)
"Sawāniḥ fī al-Lughah wa al-Muṣṭalaḥāt", MMʿIʿA, vol.xl, no.2, pp.361-8.
Shirr, Addi (1908)
"al-Alfāḍh al-Fārisiyyah al-Muʿarrabah", Beirut.
Smeaton, B. Hunter (1973)
"Lexical Expansion Due to Technical Change", Indiana.
Steingass, F. (1892)
"Persian-English Dictionary", London.
Stene, Aasta (1945)
"English Loanwords in Modern Norwegian: A Study of Linguistic Borrowing in Process", London.
Sprenger, A. (1854)
"A Dictionary of Technical Terms Used in the Sciences of the Mussalmans", Calcutta.
Sterling, R. (1904)
"A Grammar of the Arabic Language", London.
Stetkevych, Jaroslav (1970)
"The Modern Arabic Literary Language", Chicago.
al-Suwaysiyy, Muḥammad (1976)
"Mushkilat Waḍʿ al-Muṣṭalaḥ", al-Lisān al-ʿArabiyy, vol.xii, no.1, pp.9-15.
al-Suyūṭiyy, Jalāl al-Dīn (1326 AH)
"Bughyat al-Waʿāt fī Ṭabaqāt al-Lughawiyyīn wa al-Nuḥāt", checked by M.A. al-Khānjiyy, Cairo.
al-Suyūṭiyy, Jalāl al-Dīn (1438 AH)
"al-Mutawakkiliyy", Damascus.
al-Suyūṭiyy, Jalāl al-Dīn (1967)
"al-Itqān fi ʿUlūm al-Qur?ān", 4 vols, ed. Muḥammad Abū al-Faḍl Ibrāhīm, Cairo.
al-Suyūṭiyy, Jalāl al-Dīn (n.d.)
"al-Muzhir fi ʿUlūm al-Lughah wa Anwāʿiha", 2 vols, 3rd edn, eds Muḥammad Aḥmad Jād al-Mawlā et al., Cairo.
Syrian Academy (1977)
"Muʿjam al-Kīmyā?", Damascus.

al-Thaᶜālibiyy, ᶜAbd al-Malik Abū Manṣūr (1938)
"Fiqh al-Lughah", eds Muṣṭafa al-Saqqā et al., Cairo.
Ṭāha, Salīm (1976)
"al-Taᶜrīb wa Kibār al-Muᶜarribīn fī al-Islām", Sumer (Journal of Archaeology and History in the Arab World, Baghdad), vol.xxxii, pp.339-89.
al-Tahānuwiyy, Muḥammad ᶜAli (1963)
"Kashshāf Iṣṭilāḥāt al-Funūn", translated from Persian by A.N.M. Ḥasanayn and checked by L. ᶜAbd al-Badīᶜ, Cairo.
al-Ṭahṭāwiyy, Rifāᶜah (1973-74)
"al-Aᶜmāl al-Kāmilah li-Rifāᶜah Rāfiᶜ al-Ṭahṭāwiyy", 3 vols, ed. Muḥammad ᶜAmārah, Beirut.
Tarzī, Fu?ād Ḥannā (1967)
"al-Ishtiqāq", Beirut.
Taylor, Walt (n.d.)
"Arabic Words in English", SPE Tract No. XXXVII, printed in Great Britain, pp.565-600.
al-ᶜUbaydiyy, Rashīd (1979)
"al-Ḥurūf al-Mudhallaqah wa Tafāᶜuluha maᶜ al-Aswāt al-Lughawiyyah", al-Ustādh (Bulletin of the College of Education, Baghdad University), vol.ii, pp.291-327.
Ullmann, S. (1962)
"Semantics: An Introduction to the Science of Meaning", Oxford.
Wāfi, ᶜAli ᶜAbd al-Wāḥid (1972)
"Fiqh al-Lughah", 7th edn, Cairo.
Wāli, Ḥusayn (1936)
"Ism al-Alah", MMLᶜAM, vol.i, 1936, pp.371-78.
Webster's New Collegiate Dictionary (1977), Springfield, Mass.
Weinreich, Uriel (1968)
"Languages in Contact", Paris.
Wright, W. (1967)
"A Grammar of the Arabic Language", 3rd edn, Cambridge.
al-Yasūᶜiyy, Rāfā?īl Nakhlah (1960)
"Gharā?ib al-Lughah al-ᶜArabiyyah", 2nd rev. edn, Beirut.
al-Zabīdiyy, Muḥammad al-Murtaḍa (1306-7 AH)
""Tāj al-ᶜArūs", 10 vols, Cairo.
Zaydān, Jurjī (1902-6)
"Ta?rīkh al-Tammadun al-Islāmiyy", 5 vols, Cairo.
Zaydān, Jurjī (n.d.)
"Ta?rīkh Adab al-Lughah al-ᶜArabiyyah", 4 vols, Egypt.
Zaydān, Jurjī (n.d.)
"al-Lughah al-ᶜArabiyyah Kā?in Ḥayy", revised by Murād Kāmil, Cairo.
Zaydān, Jurjī (n.d.)
"al-Falsafah al-Lughawiyyah wa al-Alfāḍh al-ᶜArabiyyah", ed. M. Kamil, Cairo.
Zgusta, Ladislav (1971)
"Manual of Lexicography", Paris.

Glossary of technical terms

حالة النصب	accusative
المبنى للمعلوم (فعل)	active
ظرفي	adverbial
الاثبات	affirmative
اللاحقة	affix
القياس	analogy
الادغام/ المماثلة	arabicization
التعريب	assimilation
استعارة	borrowing
المشتركات اللفظية	cognate
مركب	complex
تصريف الافعال	conjugation
الحرف الصامت	consonant
العنقود الصوتي	consonant cluster
الصائت المركب	diphthong
المخالفة/ التخالف	dissimilation
الاشتقاق	derivation
الاسقاط/ الحذف	elision
التأثيل	etymology

gender	الجنس
gemination	التضعيف
guttural	حلقى
hollow verb	الفعل الاجوف
infix	اللاحقة الوسطية
juncture	الوصل
lexical adaptation	التكيف اللفظى
lexical development	التطور اللفظى
lexical growth	النمو اللفظى
lexical pattern	الصيغة/ الوزن
liquid sounds	حروف الذلاقة
loanform	الصيغة (اللفظية) المستعارة
mixed compound	النحت المزجى
morpheme	المورفيم/ الوحدة الصرفية
morphological patterning	الوزن الصرفى
non-native	غير اصلى (غير محلى)
paradigm	جدول التصريف
participle	اسم فاعل أو مفعول
particle	اداة
passive	المبنى للمجهول
pause	الوقف
phoneme	الفونيم/ الوحدة الصوتية
prefix	الاحقة الابتدائية
preposition	حرف جر
root (stem)	الجذر

semantic	دلالى
stress	النبرة
strong verb	الفعل السالم/ الصحيح
suffix	اللاحقة النهائية (الكاسعة)
syllabification	التقطيع
syllabic structure	بنية المقطع
syllable	المقطع
syntactic	نظمى (خاص بنظم الكلام)
syntax	النحو
transitive	(الفعل) المتعدى
weak verb	الفعل المعتل

والحضـاريـة واثرها في عملية النمو اللغوي وولادة المصطلحات العلمية على وجه التحديد ، وسمات التكيف اللفظي عبر مراحـل التطـور الاجتـماعي والعلمي ، وخصـائص اللغـة العربية وعلاقتها بالتفاعل الحضاري وموقف اللغويين العرب منها والتعريب القياسي واهميته في ضوء الحاجة القائمة الى المصطلحات والعوامل ذات الصلة بامكانية تطبيق هذا النوع من التعريب في لغة العلم ، والدور الذى تمارسه المجامع اللغوية في معالجة مشكلة النقص في المصطلحات الى غير ذلك من المواضيع ذات العلاقة .

اضافة الى ما تقدم يجد القارىء في نهاية كتابنا هذا عددا من الملاحق ادرجنا فيها مجموعات من الامثلة والايضاحات لتراكيب وظواهر لغوية مر ذكرها في مواضع مختلفة من الدراسة لقد ارتأينا وضعها في خاتمة الكتاب ابتغاء الايجاز في متنه مكتفين بالامثلة والشواهد الضرورية لتحقيق غرضنا في توضيح مانحن بصدده من وسائل وآراء .

اخيراً ، لابد لنا ان نقول : ان غاية مانطمح اليه هو ان يكون الجهد الذي بذلناه في انجاز هذه الدراسة وماتوصلنا اليه من نتائج بمستوى المكانة السامية التي تحتلها لغتنا العربية في نفوسنا وضمائرنا ، مع اعتزازنا بالفضل لكل من سبقنا في هذا الميدان .

والـلـــه ولـي الـتـوفـيق

الدكتور عبد الصاحب مهدي على
رئيس قسم الترجمة
كلية الاداب/ الجامعة المستنصرية
بغداد - العراق

دبيب الحياة في أوصاله من جديد حتى ظهرت بوادر نهضة لغوية وتعاظم الاحساس بالحاجة الملحة إلى مثل تلك النهضة لما ألم باللغة من ضرر جراء الركود الذي انتابها طيلة الحقبة المظلمة من حياة مجتمعها . فلقد وجدت اللغة العربية نفسها في صراع شديد مع لغات أجنبية تنافسها في التعبير عن مسميات الحضارة الحديثة ومفاهيمها ، وفي القدرة على النمو والتكيف السريع ومواكبة حركة التطور العلمي والحضاري . وتبين انها تعاني من نقص كبير في المصطلحات العلمية والتقنية مازال قائماً حتى في عصرنا الحاضر ، بل وبشكل متزايد ، ولا حاجة بنا إلى جهد كبير للتدليل على هذه الحقيقة فما زال عدد كبير من المؤسسات العلمية والثقافية وفي عدة أقطار عربية تستعين باللغة الانجليزية أو الفرنسية مصادر لاستعارة الألفاظ والمصطلحات أو لغة للتعليم والبحث والتعبير عن الأفكار .

ان مشكلة المصطلحات ونمو اللغة العربية العلمية بشكل عام تنطوي على جوانب عدت منها ما يتعلق بطبيعة اللغة وخصائصها الذاتية ومنها ما يتمثل بمعايير وسمات تميز لغة العلم عن لغة الادب . هذا فضلا عن عوامل اخرى خارجة عن اللغة ولكن تؤثر فيها كالحضارة والتراث والمجتمع وغيرها مما سيرد ذكره في الفصول اللاحقة من دراستنا .

ان هذه المشكلة رغم اهميتها وكونها حظيت باهتمام العديد من الباحثين والمؤسسات اللغوية المتخصصة والهيئات الحكومية ، فانها مازالت تتطلب دراسة موضوعية تتبنى المنهجية العلمية في البحث وتتخذ من الحقائق التى توصلت اليها الدراسات اللغوية الحديثة معالم اساسية تستهدى بها للوصول الى النتائج المنشودة ، فقد جرت محاولات عده للبحث في بعض جوانب هذه المشكلة كما سنبين لاحقا - الا انها للاسف لم تكن عند المستوى المطلوب من الدقة والكفاية والموضوعية لذلك فقد حاولنا في دراستنا هذه ان نتجنب المآخذ المشار اليها لعلنا نبلغ بها المستوى الذى نطمح اليه املين ان توفر لنا هذه المحاولة فرصة افضل لالقاء الضوء على واقع المشكلات اللغوية للعربية وبذلك نكون اكثر قدرة على تفهمها ومعالجتها .

ان من بين الاهداف الرئيسة التي نهدف الى تحقيقها في هذا الكتاب هو دراسة نمو وتطور المصطلحات العلمية والتقنية في اللغة العربية عبر مراحلها التاريخية المختلفة ، مع التركيز بشكل خاص على المرحلة الحاضرة كما نهدف الى دراسة وتقويم قابلية العربية على الاستجابة للمطالب المتزايده التى تتمخض عنها الحضارة الانسانية بمختلف جوانبها العلمية والتقنية والفكرية والاجتماعية ، اضافة الى امكانياتها الذاتيه وقدراتها الكامنة على اغناء نفسها بنفسها ووسائلها الطبيعية في النمو والتطور . وعلى صعيد اخر عمدنا الى اجراء دراسة تحليلية نقدية للعديد من الآراء والمقترحات التي تقدم بها عدد كبير من الكتاب واللغويين والمعجميين واعضاء المجامع اللغوية وغيرهم في محاولات من جانبهم لايجاد ما يمثل بالنسبة لهم حلولا مناسبة لمشاكل أو حاجات مصطلحية معينة . ولاجل اقامة الدليل على راينا ازاء هذه المحاولات أو غيرها مما يمثل وجهة نظرنا الخاصة فقد عمدنا في البحث الى استطلاع الواقع اللغوي الحقيقي أو الاستعمال الفعلي للغة في مجالات استخدامها المتنوعة مع عدم اغفالنا للجوانب النظرية ذات العلاقة . اضافة الى ذلك ، وفي حالات كثيرة وجدنا من المناسب ان نعزز مانورده من آراء بالاشارة الى مايحصل في لغات اخرى ، حيث تكون الظاهرة المطروحة للنقاش والمتعلقة باللغة العربية لها ما يماثلها في تلك اللغات .

تتألف دراستنا من خمسة فصول ، تناولنا في الاول منها بنية الكلمة العربية واردنا هنا ان نعرف القارىء بالقواعد الاساسية التي تستند اليها عملية الاشتقاق أو الصياغة اللفظية في اللغة العربية . وتضمن هذا الفصل ايضا تحديداً لجملة من المصطلحات والمفاهيم التي يتكرر استخدامها في الدراسة .

اما الفصول الثاني والثالث والرابع فتشتمل على دراسة تفصيلية لثلاث طرق رئيسية لايجاد المصطلحات واهميتها النسبية في بناء اللغة العربية العلمية . وهذه الطرق هي : الاشتقاق والنحت والتعريب . وقد عنينا هنا بالدور الذى تضطلع به كل من هذه الطرق وما يتعلق بتطبيقاتها العملية ومناسبتها للغة العلمية ومدى كفايتها .

هذا اضافة الى ما توصلنا اليه من نتائج ومقترحات للنهوض باللغة العربية العلمية واثرائها .

واما الفصل الخامس فقد ضمناه استطلاعا اجريناه حول عينات من المصطلحات العلمية والتقنية والفاظ الحضارة في اللغة العربية وقد اجرينا هذا الاستطلاع في عدد من كليات جامعة بغداد وفي اقسام متنوعة فيها . اوضحنا في بداية هذا الفصل الاسباب والغايات التي دفعتنا الى القيام بهذا الاستطلاع ، ثم ذكرنا النتائج التي توصلنا اليها . وهناك ايضا جملة استنتاجات تتعلق بالامكانية المستثمرة وغير المستثمرة لانواع مختلفة من وسائل الصياغة اللفظية اضافة الى عدد من العوامل المتعلقة بقبول أو انتشار الالفاظ والمصطلحات الجديدة .

ان ما انجز في هذا الفصل يمثل اختبارا عمليا لعدد من المسائل المصطلحية المثيرة للجدل التي سبقت دراستها في اطار نظري نسبيا ، في الفصول ٢ - ٤ .

اردنا مما ورد ذكره اعطاء القارىء الكريم فكرة موجزة عن اهم الاهداف التي توخيناها في هذا الكتاب . وهناك مسائل اخرى كان لها نصيب كبير من اهتمامنا لعلاقتها بموضوع الدراسة ومن بين تلك المسائل : العوامل التاريخية

مقدمة المؤلف

لا يخفي على القارىء الكريم ما للغة من أهمية كبرى في حياة المجتمعات البشرية وما تضطلع به من مهمات أساسية في عمليات التفاعل الانساني والتواصل الفكري والترافد الحضاري .

ومع ان هذه الحقيقة تنطبق على المجتمع البشري في مراحل تطوره المختلفة ، فانها تبرز بشكل واضح في العصر الحديث حيث بلغت الحضارة درجة كبيرة من النمو والتعقيد وبخاصة في ميادين العلم والتقنية . وبما ان حياة اللغة وديمومتها تقوم على مواكبتها لروح العصر أو المجتمع الذي تعيش فيه ، شأنها في ذلك شأن أي كائن اجتماعي فلابد لها من الاستجابة لحاجات عصرها فتكون بذلك اداة طيعة وقادرة على التعبير عن شتى المفاهيم الحضارية والعلمية وبدرجة من الكفاءة والدقة توازي تلك التي تتسم بها العلوم الحديثة في منهجيتها وطرق بحثها ومنجزاتها . وليس في هذا القول اشارة إلى لغة معنية دون غيرها أو مجتمع دون سواه ، وانما نحن بصدد ناموس تخضع له كل لغات البشر وسمة طبيعية يولدها التفاعل الذي لابد منه بين اللغة والمجتمع أو بين اللغات والمجتمعات البشرية بشكل عام . ومصداقاً لكلامنا هذا فعند دراستنا لحركة النمو والتطور في لغة من اللغات وعبر مراحل تاريخها المختلفة ، نجد ان هذه الحركة قد تتفاوت أو تتباين من مرحلة إلى أخرى ، فقد تكون نشطة وسريعة وغنية في مرحلة معينة ، بينما تكون خاملة وبطيئة وفقيرة في أخرى . ولودرسنا الجوانب الاخرى لحياة المجتمع المعني وحركة نموه ، وفي المراحل ذاتها ، لوجدنا الصورة نفسها ولرأينا ان عملية النمو في اللغة تمثل جانباً واحداً من جملة الجوانب المختلفة التي تشكل الحركة العامة لتطور المجتمع ونموه .

ولنا في لغتنا العربية الفصيحة مثال واضح على ماتقدم ذكره . فلقد مرت هذه اللغة عبر تاريخها بمراحل مختلفة تفاوتت فيها درجة نموها وطبيعة تكيفها . فنجدها قبل الاسلام لغة شعر وخطابة وتداول مسموع بالدرجة الأولى ، غير انها بعد ذلك ونتيجة للقيم والمفاهيم الجديدة التي أوجدها الدين الحنيف والتحولات الاجتماعية والفكرية التي طرأت في حياة المجتمع الجديد ، مالبثت ان مهدت لنفسها سبيل التكيف للمناخ الجديد فولدت الكثير من المفردات والمصطلحات وكيفت قسماً آخر مما لديها للتعبير عن الافكار والمفاهيم الجديدة والمتزايدة في مختلف الميادين . فكانت الفصيحة بذلك لغة الفقه والشريعة والفلسفة وغير ذلك من المواضيع التي نمت وازدهرت في تلك الفترة . وحين تم الفتح الاسلامي لكثير من المناطق استطاعت اللغة العربية ان تثبت كفايتها في مثل تلك الظروف ، فاذا بها لغة الادارة والقضاء والسياسة ثم جاءت مرحلة النهضة العلمية التي بلغت أوجها في العصر العباسي ، فنشطت حركة ترجمة العلوم بشكل لم يسبق له مثيل حيث كان بيت الحكمة - تلك المؤسسة الاكاديمية - زاخر بالعلماء والمترجمين وحافلاً بالمؤلفات والتراجم ووسائل البحث العلمي . فكانت تلك المرحلة والحالة هذه ، غنية بالمنجزات العلمية ، واضحة الأثر في صورة المجتمع الذي عاصرها وطبيعة الحالة التي اكتنفتها . ومرة أخرى نجد اللغة العربية وقد حثت خطاها لتواكب المسيرة العامة لمجتمعها خطوة بخطوة . فاذا بقوتها التعبيرية تتعدى حدود المفاهيم الادبية والعلوم الشرعية لتشمل ميادين أخرى في العلوم الطبيعية والتطبيقية كالطب والكيمياء والهندسة والفلك وغيرها . وإذا بها تكتسب قدرة عالية على الدلالة الدقيقة والواضحة استجابة لتطلبات لغة العلم . ثم نجد انها - اضافة إلى استثمارها لقدراتها الذاتية وانتهالها من مواردها الاساسية - لا تنغلق على نفسها ولا تتقوقع في مستقر العزلة والاختناق ، وانما تحرص على مواصلة الحياة من استجابتها لنواميس التطور اللغوي المتزامنة مع التطور الحضاري أو الناتجة عنه ، مع تمسكها الشديد بجوهر شخصيتها ولهذا فهي لم تجد ضيراً في الاستفادة من الموارد غير الذاتية - تلك التي تلجىء اليها الضرورة القصوى لاعتبارات لغوية وحضارية سنوردها ضمن فصول هذا الكتاب .

ثم تلت ذلك مرحلة أخرى امتدت منذ نهاية القرن الثالث عشر وحتى القرن التاسع عشر بدت اللغة فيها وكأنها تمر في حالة استقرار فقد تباطأ نبضها وضعف نشاطها . ولو تأملنا الأمر لوجدنا ان سبب ذلك يكمن في حالة الركود العام والشلل الذي أصاب الحياة في المجتمع العربي نتيجة للغزوات الكثيرة التي تكالبت على الأمة والهجمات الشرسة التي تعرضت لها حضارتها في تلك الفترة المظلمة ولعل أوضح مثال على ذلك ان بغداد التي كانت تعرف انذاك بمدينة السلام والتي ذاع صيتها بصفتها مركزاً مشعاً للحضارة والنشاط العلمي ، تعرضت سنة ١٢٥٨ لهجمات المغول الذين احلوا فيها الدمار والخراب واطفأوا في ربوعها شعلة الفكر والحضارة ، وقد انعكس ذلك على اللغة فأصابها ماأصاب مجتمعها من الخمول وتباطؤ النمو .

وفي فترة الاحتلال العثماني والاستعماري الغربي للوطن العربي حرمت اللغة العربية من حقها في ان تكون لغة التعليم في المدارس أو اللغة الرسمية في الدوائر . وكان لحالة الشلل التي عاشها المجتمع انذاك أثرها في انحسار حركة الابداع الفكري والبحث العلمي . وبما ان ميلاد المصطلحات العلمية يكون عادة على ايدي العلماء والباحثين وفي مختبراتهم ، فان غياب مثل هذه الفئة من صناع اللغة كان احد العوامل المهمة في تعطيل عملية النمو اللغوي في تلك الحقبة .

غير ان اللغة العربية وعلى الرغم من الظروف التي سلف ذكرها ظلت تنتظر اللحظة التي ينهض فيها المجتمع العربي من كبوته لتواصل مسيرتها معه من جديد . ولقد كان ذلك حين ادرك العرب انهم يقفون وجهاً لوجه أمام تحديات الحضارة الاوروبية . فكانت تلك الوقفة بمثابة هزة عنيفة منحتهم الفرصة ليتأملوا فيما آلت إليه أمورهم وليذكروا انفسهم بماضيهم المشرق وحضارة أمتهم العريقة التي ساهمت في بناء الانسانية بشكل فعال ولم تعش عالة عليها . وما ان عاود المجتمع العربي نشاطه الحضاري ودب

المستقبل . ولقد استنبط نتائجه من تحر نظري لموضوعه ، والتشاور مع المتخصصين بهذا الفرع من المعرفة ، وبالشرح والتحليـل للمفـردات والمصطلحـات الشـائعـة الاستعـمال وكـذلك بالاطلاع على القواميس والدوريات التي تتناول موضوعات في هذا المجال . وقد أجرى الدكتور« علي» بحثاً ميدانياً على طلبة يدرسون مواد علمية بهدف اختبار مدى استعدادهم لتقبل مصطلحات جديدة ثم اعادة صياغتها باستخدام هذه الوسائل المختلفة ، وكذلك مدى السهولة التي يمكن بها أن يفهموا هذه المصطلحات .

وقـد تنـاول الـدكتـور «عـلي» بالبحث أيضـاً الـوسائل المتعددة التي استخدمت من قبل لتكوين الكلمات ومن هذه الـوسـائـل «المـزج» ، و «الدمج» و «التركيب» . مع تسجيل رد فعل الطلبة عند سماع أو قراءة الكلمات التي استنبطت بأستخدام احدى هذه الوسائل .

وتمثل الدراسة التي قام بها الدكتور «علي» مرحلة جديدة في طريقة البحث العلمي للعربية تجمع بين احترام للتقاليد اللغوية مبني على علم تام باللغة وبأصول الكتابة المتعارف عليها في مجال اللغويات مقروناً بطريقة بحث رحبة الصدر لمعالجة المشاكل الحقيقية التي يواجهها الطلاب ، وبتغطية جيدة للطريقة التي تستخدم في الكتابة التقنية في الوقت الحاضر .

المحَرر
بروس انغام

تقديم

ان موضوع هذا الكتاب عزيز على أذهان العلماء العرب ، وهو في الوقت ذاته يثير أموراً تختلف تماماً عما تعود عليه القارىء الانجليزي .

تتعلق هذه الأمور بخلفية لغوية تتميز إلى حد ما بالميل إلى قبول مفردات غريبة عنها ، فالانجليزية لغة مطعّمة - وهي أساساً لغة جرمانية مرّ عليها تسعمائة عام من التشرب بالمفرادت الرومانية وقد بدأ ذلك أول مابدأ عن طريق التأثير بالفرنسية النورماندية ، وبعد ذلك كان نتيجة لعمليات الاقتباس الأدبية من اللاتينية . ولكونها لغة مطعّمة - كما ذكرنا - فالانجليزية تحتوي على كم هائل من المفردات الأجنبية والتي تعتبر وبلا أي حرج - انجليزية بحتة .

والعربية - من ناحية أخرى - في وضع مختلف تماماً اذ ليس لأي لغة أن تدّعى الانسجام التام في مفرداتها كما يحق للعربية أن تدّعى . ومن الصحيح أن الجزء الأكبر من مفرداتها الأساسية اما ذا أصل عربي أودخل اللغة نتيجة لعمليات الاقتباس من الثقافات القديمة ، وأمتصته اللغة إلى درجة لايمكن معها تمييزه عن مفرادتها الأصلية .

والحقيقة التي لاتنكر أن اللهجات المتعددة قد تشربت الكثير من مفردات الثقافات الأجنبية غربيها وشرقيها على السواء ، تلك الثقافات التي كانت هي على صلة بها ، بيد أن المؤسسات الأدبية في العالم العربي قد قاومت دائماً دخول هذه المفردات إلى اللغة المكتوبة ، وقد ظهرت الحاجة إلى نقل المفاهيم العلمية الحديثة إلى العربية في ميادين كثيرة خاصة بعد حدوث تطورات تربوية وتقنية في شتى أرجاء العالم العربي .

وقد شعر بعض العلماء أن الوسيلة الوحيدة المقبولة لنقل هذا إلى حيز التنفيذ هو توسيع معاني الكلمات العربية المستخدمة أو صياغة كلمات جديدة عن طريق الاشتقاق من الأصول الموجودة . ويعرف هذا الأسلوب بالقياس الصرفي «الاشتقاق بالتكوين المتماثل» . واتجه آخرون إلى تبني المفردات الأجنبية مع تعديلها لتناسب التراكيب الصوتية والصرفية للغة العربية . ان أمكن إلى ذلك سبيلاً ، ويعرف هذا «بالتعريب» .

وللعرب الحق أن يفخروا بغنى لغتهم بالمفردات وبالعدد الكثير من المترادفات والكلمات التي تعبر عن أدق الاختلافات في المعنى .

وقد شعر أنصار القياس الصرفي أن وجود مثل هذا الكم الهائل من جذور الكلمات مقروناً بقواعد صرفية خاصة باللغات السامية والتي تتمثل في تكوين المفردات عن طريق اضافة سوابق أو لواحق لأصل الكلمات - وجدوا أنه من الممكن - وبقدر معين من الأصالة - أن يتم التكيف مع الحاجة الملحة لادخال مفردات جديدة في اللغة . علاوة على هذا ، فقد وجد أنه من الممكن استخدام الكلمات المهجورة والتي لم تعد تستعمل على نطاق واسع لصياغة مفاهيم جديدة .

ومن جهة أخرى فأن أشياع التعريب انتصروا للرأي القائل بأنه بالرغم من أن تكوين الكلمات من أصول عربية هو الأفضل عند امكانية ذلك ، الأ أن غموض بعض هذه المصطلحات الجديدة كان شديداً لدرجة أنه وجد أن استخدام الكلمات الأجنبية يوفر الكثير من الجهد بالاضافة إلى أنه له ميزة الألفة والاعتياد بين أولئك الذين هم دراية بلغات أجنبية مما يؤدي إلى ايجاد صلة بين الثقافات الغربية والشرقية .

ويتناول الدكتور «علي» بالبحث هاتين الوسيلتين ويحاول أن يضع خطوطاً عريضة لأي تطور ممكن حدوثه في

بسم الله الرحمن الرحيم

تأليف
الدكتور عبد الصاحب مهدي علي
رئيس قسم الترجمة
كلية الآداب - الجامعة المستنصرية
بغداد - العراق

دراسة لغوية
عن تطور المصطلحات العلمية
في اللغة العربية

الكتاب السادس

مؤسسة كيغان بول العالمية
لندن - هنلي - بوسطن
١٤٠٦ هـ - ١٩٨٦ م

مكتبة اللسانيات العربية
سلسلة كتب عالمية في الدراسات اللغوية العربية

دراسة لغوية
عن تطور المصطلحات العلمية
في اللغة العربية

مكتبة اللسانيات العربية

الحمد لله وحده ، والصلاة والسلام على من لا نبي بعده . أما بعد : فإن هنالك أسباباً عدة دعت إلى إنشاء هذه السلسلة من الكتب في حقل اللسانيات والصوتيات العربية .

أولاً : إن هذا الحقل يمر بتطور سريع في إطار الدراسات اللغوية المعاصرة . كما أن كثيراً من الجامعات العربية والغربية قد بدأت تدخل علم اللسانيات وعلم الصوتيات وبعض العلوم اللغوية الحديثة ضمن مواد التدريس بها . بالإضافة الى الإهتمام المتزايد في الدوائر اللغوية العالمية بهذا الميدان .

ثانياً : ومع ازدياد الاهتمام بالدراسات اللسانية والصوتية العربية بدأت تصل هذه الدراسات إلى مرحلة متقدمة في النضوج ليست مستفيدة من معطيات علم اللسانيات العام والعلوم الأخرى النسبية فحسب ، بل وأيضاً من معطيات الدراسات اللغوية العربية القديمة .

ثالثاً : بدأت تظهر في حقل اللسانيات العربية فروع ونظريات مختلفة تشمل الصوتيات والفونولوجيا والنحو والدلالة ، وعلم اللغة النفسي ، وعلم اللغة الاجتماعي ، وعلم اللهجات العربية ، وصناعة المعاجم ، ودراسة المفردات ، وتدريس العربية أو تعلمها كلغة أولى أو ثانية أو أجنبية ، وعلم الاتصال ، وعلم الإشارات اللغوي ، ودراسة المصطلحات ، والترجمة ، والترجمة الآلية ، وعلم اللغة الإحصائي ، وعلم اللغة الرياضي ، وتاريخ العلوم الغوية العربية ، وما إلى ذلك .

يضاف إلى هذا كله أن الاقبال على اللغة العربية دراسة وتدريساً وبحثاً يزداد يوماً بعد يوم على الصعيدين المحلي والدولي . ولما لم يكن هناك منبر يرتفع منه نداء لغة الضاد وتعلو منه أصوات الباحثين والمتخصصين فيها لذا وجدت مكتبة «اللسانيات العربية» لتسد هذا الفراغ الكبير والفجوة العميقة وتدفع بالبحث اللغوي العربي قدماً إلى الأمام خدمة للغة القرآن الكريم والتراث العربي الأصيل ، وسيراً بالبحث اللساني العربي للحاق بركب اللسانيات العامة المتقدم ، وإثراءً للدراسات اللغوية واللسانية العربية .

وتحرص هذه السلسلة العالميّة على تقديم الجديد من البحث اللغوي وإعطاء الفرصة للبحاثين من العرب وغيرهم للمشاركة في بناء صرح اللسانيات العربية حتى تستعيد الدراسات اللغوية مجدها الماضي العريق .

ولأن هذه السلسلة تعد الأولى من نوعها في الدراسات اللسانية العربية المتخصصة ، فإننا نهيب بكل باحث متخصص في مجال اللسانيات العربية بمختلف فروعها النظرية منها والتطبيقية أن يشارك بجهوده وأفكاره وأبحاثه وألا ينجل بتقديم أجود ما لديه من عطاء في سبيل دعم أهداف هذه السلسلة وتطوير مجالاتها الواسعة . والباب مفتوح أمام جميع الأقلام العربية والشرقية والغربية التي تخدم هذه الأهداف الخيّرة .

ونسأل الله العلي القدير أن يحقق لهذه السلسلة ما تصبو اليه من نجاح وتقدم . قال سبحانه وتعالى :

«وقل اعملوا فسيرى الله عملكم» ، صدق الله العظيم . والله الموفق لما فيه الخير والصواب لصالح أمتنا العربية الإسلامية المجيدة ولغتها العريقة الأصيلة . إنه سميع مجيب .